# TECHNOLOGY ASSESSMENT AND DEVELOPMENT

# TECHNOLOGY ASSESSMENT AND DEVELOPMENT


**Edited by**

**Mangalam Srinivasan**

*Center for*
*International Technical Cooperation*
*American University, Washington, D.C.*


**Foreword by**

**Hon. Indira Gandhi**

*Prime Minister of India*


PRAEGER

PRAEGER SPECIAL STUDIES • PRAEGER SCIENTIFIC

**Library of Congress Cataloging in Publication Data**
Main entry under title:

Technology assessment and development.

Includes index.
1. Technology assessment—Addresses, essays, lectures. 2. Underdeveloped areas—Technology—Addresses, essays, lectures. I. Srinivasan, Mangalam.
T174.5.T415 338.9'009172'4 82-3795
ISBN 0-03-059543-6 AACR2

Grateful acknowledgment is hereby made to City Lights Books for permission to use the following copyrighted material: Howl and Other Poems by Allen Ginsberg (San Francisco: City Lights Books, 1959), p. 17. Copyright © 1956, 1959 by Allen Ginsberg. Reprinted by permission of City Light Books.

While the linguistic inadequacies of the English language have imposed upon me the use of "he," "him," and "his" in general reference to a human being, please note that these words also refer to that type of man known as woman.

Published in 1982 by Praeger Publishers
CBS Educational and Professional Publishing
a Division of CBS Inc.
521 Fifth Avenue, New York, New York 10175 U.S.A.



123456789 052 987654321

Printed in the United States of America

# FOREWORD
## Hon. Indira Gandhi

A human being's development is influenced by biological as well as environmental factors. In any given person there are traits that can be attributed to genetic inheritance. Other characteristics, such as manner of speech and cultural preferences, may be largely fashioned by his social milieu.

Laws of physics and chemistry, long considered predictable and set, are being given a fresh look. Advancing research indicates the precariousness of their transitory versions. The life sciences are more bewildering. There is a core of science over which we have no control. Technology, however, is different. It is conditioned by the society it serves, for it is the application of science to achieve a specific human purpose, the choice of which depends on existing circumstances as well as the objectives to be accomplished. In that sense all technology is "appropriate." Technology for technology's sake makes little sense.

A realistic assessment of a country's potential and aspirations and the adoption or adaptation of available technology are important steps in scientific planning. One cannot be dogmatic in such matters. One has to persevere to broaden the horizons of knowledge.

The development and developing economies of our times are good material for research on the impact of science and technology on contemporary people. Many expensive mistakes have been made by various countries in their dealings with technology. Hiroshima and Nagasaki will remain colossal and grim reminders of man's insanity. At the same time, developing countries have considerable opportunity to go ahead with a clearer perception of how to control technology and make proper use of it.

Developing countries have an ethos of traditional beliefs and are also closer to the natural and the spontaneous. Getting technology that is apparently inconsistent with this approach has been necessarily slow. Nor has the attitude of the affluent countries, which were willing to share technological advances only on their own terms, helped the process of modernization. Only too often has technological assessment operated in a seller's market. Environmental considerations that for centuries had been paramount in the practice of traditional technologies became blurred in the race for agricultural industrialization and commercialization. Desirable in themselves, these phenomena were not harmonized in the strategy of balanced development, except in a framework of long-term perspective planning. India, for instance, took care of some factors and has now built a base where technology assessment has become a useful and necessary condition for further progress.

By bringing together the experiences of China, India, Mexico, the United States, the USSR, and Yugoslavia, among others, the editor of this volume has provided rich fare for the technical as well as the lay reader. The studies on telecommunications in the development process, of management as a factor in technology choice, of the feasibility of employing used industrial machinery in less developed countries, and of integrating women in technological progress, to cite a few, deal with themes that figure constantly in the vocabulary and actions of modern governments. Informed discussion will pave the way to more judicious approaches to technology.

New Delhi
March 2, 1982

# ACKNOWLEDGMENTS

This book is the result of considerable effort on the part of the contributors, whose articles are insightful and scholarly. We thank them for their participation.

The foreword to this book comes from a distinguished world leader, Madam Indira Gandhi, the prime minister of India, whose vision, intuition, and knowledge are among the most important factors in charting the social and economic course of development in India through the use of science and technology. In part, the policies of Mrs. Gandhi's government are responsible for the position the country holds as a nation, with the third most scientifically and technically trained work force in the world. This massive surge in scientific and technological achievements notwithstanding, India faces some of the most severe problems of any developing nation; hence, the special interest in Madam Gandhi's message regarding the interplay of technology assessment and human development.

My sincere thanks are owed to Rodney Douglas, my research assistant, who offered valuable and wise editorial assistance and suggestions on the organization of the book. He worked diligently, joyfully, and with dispatch toward the completion of the book.

I owe a special debt to Ramachandran Krishnan, whose editorial and research assistance contributed greatly toward the accomplishment of this effort.

I would also like to express my grateful appreciation to Frank Barrow and the Center for International Technical Cooperation at the American University, without whose enthusiasm and support this book would not have been possible.

# CONTENTS

Chapter | Page

LIST OF TABLES

# INTRODUCTION

> It is with our passions as it is with fire and water; they are good servants, but bad masters.
>
> Aesop

> "Well, in our country," said Alice, still panting a little, "you'd generally get to somewhere else—if you ran very fast for a long time as we've been doing."
>
> "A slow sort of country!" said the Queen. "Now, here you see it takes all the running you can do to keep in the same place. If you want to get somewhere else, you must run at least twice as fast as that!"
>
> Lewis Carroll
> Alice in Wonderland

The implications of technology assessment (TA) for policy for the economic and social reconstruction of a developing nation are just beginning to be understood. Since the decade of the 1940s, when many countries started to free themselves from the old colonial powers, until most recently, transfer of "the package" called technology meant freedom from poverty, ignorance, and ill health. It also meant wealth and power. However, decades after the arrival of and occupation by modern Western technologies, most of these countries seem to be running only to find themselves in the same place from which they started.

Despite the great strides that have been made to alleviate drudge labor, increase productivity, and enhance the quality of life through scientific and technological advancements for the majority of the world's people, material conditions of life have worsened, while alienation and despondency have extended their domain to every level of human existence. Among the reasons contributing to this situation are lack of political commitment, increased expectations, too much growth too fast (regardless of the impact on values and societal goals), and irrelevant and irrational choices of technology. Considerations of relevance have often been sidestepped by what are politically expedient sciences and internationally prestigious technologies. In many instances, even when the choice of technology was guided by rational policies, technological impacts have often exceeded the expectations and intended consequences, thus greatly upsetting all balances: social, economic, and ecological. These deliberations have moved technology assessment efforts close to political power and decision-making circles in many countries.

National science policy is a matter of political choice and goal. As such, it has been argued that technology assessment is a mere ritual that ratifies the policy already in existence. Technology assessment, it has been held, does nothing more than merely "affirm the consequent." So why all the fuss? Why can technology not be assessed as a matter of regular policy analysis? Technological change, although only one of many changes, is by far the most important change for accelerating the attainment of social and economic development goals. Many of today's technologies alter human ecology not only fundamentally but irrevocably. Modern high technologies offer speed, convenience, accuracy, and mass production but they also carry the potential to harm as never before. The complexities of modern technologies have demanded extraordinary measures and methods of analysis, care, supervision, and management.

Traditional policy analyses are unable to cope with the complexities of technological change and development and their significance for human development. This means that the existing apparatus needs to be strengthened using new approaches to learning, projections, assessment of benefits and risks, and understanding of the impact of proposed and alternative technologies. Typically, TA aggregates the various parameters (such as the desirabilities, uses, costs and benefits, risks and consequences, and other socioeconomic impacts) and then tests, ranks, and explicates the measure of alignment among the various parameters as they relate to the various technological options.

Policy analysis is a method by which relevant knowledge is isolated from information; important variables are identified, analyzed, and conserved; and decisions are evaluated for uncertainties and risks. Technology assessment, although a branch of policy analysis, is by far the most transdisciplinary principle and method; it is at once normative, intuitive, descriptive, synergistic, and sympodial. At the primary level, TA is technical, design oriented, and involves cost-benefit and requirement analysis; at the secondary instance it is derivative, convergent, and incidental; at the tertiary order it is didactic, intuitive, and normative, dealing with what Alvin Weinberg would term as "transcientific" considerations. At this stage, it is also a matter of consciousness, directly related to the interplay of knowledge, power, ecology, culture, and ethics with technologically mediated growth, development, and decay of human life.

In the immediate day-to-day sense, technology assessment is a delimiting, discriminating, and classifying doctrine that acts as a corollary to policy analysis and provides the critical imperatives needed to reduce the knowledge gap between policy goals and actual consequences. Unlike evaluation studies where analyses merely compare promises with performance, assessment studies are policy generating and decision oriented. Assessments afford the chance to bridge the power gap between promise and performance, to consider input-output relationships between technological choices, and to relate them to cultural interactions and impacts. Assessment verdicts often lead to social and cultural rehabilitation of technologies, while evaluations as ex post facto studies are unable to bring about corrective changes. Both processes heighten our sensitivities and enhance our consciousness with regard to technological futures. In an operational sense, both policy analysis and evaluation research can start with certain standard questions and follow standard paths of analysis. TA, on the other hand, does not start with a goal and, therefore, has no point of convergence to reach, makes no assumption as to the answers, and has no set questions. The particular technology, the socioeconomic context, and associated geopolitical considerations determine the questions.

The future is not a point radically unconnected to the past and present. The future is a steadily evolving, organic, and dynamic state with embryonic connections both to the past and present. The future is just one point, place, and segment in a continuum of time that accommodates the past, present, and beyond simultaneously. The lack of understanding of this fact is the reason for the wholesale colonization of human future by means of technology. For this reason, the evaluation of devices, methods, and techniques and the reflection on the nature and meaning of technological concepts and social assumptions of the past and present are integral parts of technology assessment. Major conceptual problems in past assessments have been in defining our technological future in the context of the people for whom and the purpose for which the technology is intended. TA has long conceded that evaluation of the purpose belongs in the realms of politics and ethics. This attitude needs to be revised, and it is to be noted that, in fact, the tension between know-how and know-what can be effectively resolved by TA, which is concerned with purposes as well. After all, we have known for some time that "there is one quality more important than 'know-how.' . . . This is 'know-what' by which we determine not only how to accomplish our purposes, but what our purposes are to be."[1] The word assess is a verb, and it is a statement of value. It also means "to sit by" when the meaning is derived from the Latin root assideo. The assessor, according to Webster, is "one who sits by another, as next in dignity; an official associate." Although in times past technology assessment has been a somewhat passive ritual and a terminal activity, increasingly it is being approached as a process and as a statement of value. "Sitting by another" does not mean sitting passively anymore; rather it means active participation in matters of decision making and decision implementation.

The relevance of all technology assessment is for policy to perform as intended in translating political goals into tangible benefits. Policy analysis is the overall canvas wherein technology assessment is but one theme. The aesthetic enthusiasm for the picture relies heavily not only upon the various individual themes attaining perfection but also on their relationship between each other and in the overall context. It is entirely possible that policy analysis can set out to examine and set forth the assumptions and define the problem focus, area of activity, scope, time, context, and purpose of the intended technology in the usual way in which policy reacts to crises. The assessment of technology must take place in a more deliberate manner and with an absence of excitement. It is not a response to crises. Although in lay terms technology has merely meant the hardware, it is just one aspect of cosmic considerations governing the relationship of technology to the future of humanity. Robert Holt expresses these ideas in the following fashion:

> The starting point of technology assessment would not be some piece of hardware that happens to be on the scene; rather, the starting point would be an analysis of the mechanisms through which social, economic and political forces manifest themselves.[2]

For these reasons, technology assessment has come to be a separate and continuous process affecting technological decisions and implementation. The principle of separation of the two activities—technology assessment and policy analysis—is one of purpose and not of essence. The functional division exists to achieve maximum benefits owing to technology for the target group. Since technology assessment is not charged with the responsibility of decision making and is essentially a didactic activity, it has been assumed that it is unbiased and neutral. Some have argued further that policy analysis is too close to the tree to see the forest, whereas technology assessment can, without penalty, spell out the alpha and omega of the technology under review. But is this true? In reality, bias has been found to permeate all realms of science, from conception to theory and from blueprint to product, quite effectively and with devastating consequences. It is true that TA does not set goals, make decisions, or work with political expediencies; therefore it can maintain some emotional distance from the problem at hand. But these are the same elements that introduce arrogance based on assumed neutrality and a lack of commitment to higher-order purposes, namely, social and economic justice and equity.

## ASSESSING TECHNOLOGY ASSESSMENT

Politicians, it is claimed, often hold scientists as hostages through funding mechanisms and political directions. In recent history, however, especially in the context of many developing nations, scientists are in a position to intimidate both the politicians and the bureaucrats. The formidable nature of the scientific establishment and the esoteric connotation afforded to the words science and scientist are often instrumental in imposing culturally and philosophically incompatible sciences and socially irrelevant technologies on many nations. The assessor cannot be taken for granted but should be carefully evaluated. Developing "tools" to judge "tools" should become an integral and necessary part of assessment activities. Standardizing assessments has been a major operational problem because the human and ecological contexts vary tremendously along with scientific notions and technological implements. There are no predestined grooves that, when in place, are capable of launching technological developments in developing countries.

The analytical imperatives in technology assessments have suffered from the same narrow vision that permeates the field of development planning. Econometric planning advocated by international development agencies has advanced heavily numerically oriented evaluations, assessments, and planning while taking a skeletal approach toward examination of who the people involved really are, what their notions may be, and what their aspirations hold for their cultural, social, and economic enhancement.

Vision and intuition are outside the scope of assessment, or so we are told. This has resulted in technological development and planning, which are directly contrary to the integrative nature of human existence. Intuition, simply stated, is merely a matter of everyday common sense and imagination. Technological progress is not separate from human progress, which in itself is a function of a variety of considerations, chief among them human vision and intuition. The assessor then has an obligation to discuss the technology as it relates not only to the various aspects, namely technical, socioeconomic, and political contexts, but also in terms of what our vision is for the future. The assessor also has the obligation to recognize that those who read his assessments deserve a full account of the shape of things to come. The assessor, of course, runs the risk of ridicule from peers by putting imagination, feelings, and even personal idioms alongside cold numbers, calculations, and equations. The truth is that such an approach greatly strengthens rather than dilutes the numbers and the technical aspects of technological choice, transfer, operation, and maintenance.

If this integrative approach is ignored or sidestepped, the result is that we have a technology assessor who matches the description of Kierkegaard's philistine.

> Devoid of imagination, as the Philistine always is, he lives in a certain trivial province of experience as to how things go, what is possible, what usually occurs . . . for the Philistine thinks that he is in control of possibility. . . . It carries possibility around like a prisoner in the cage of the probable [and] shows it off.[3]

The objective of this volume is to expose the reader to samples of assessments by different experts, some of whom have relied upon their academic disciplines, operational experiences, and intuitive and cultivated insights, while others have chosen methods of assessment rooted in the political directions and cultural goals of their respective societies. Examples of the latter method are to be found in Cheng Xuemin's account of China's hydropower development, Momir Cecez's account of the role of fixed assets in the development of Yugoslavia, and Valentina Tereshkova's essay on women in Soviet development.

This volume has been organized into three sections. Part I explores the various concepts and trends, both current and evolving, in technology assessment. Part II presents technology assessments relevant to particular countries as provided to us by a number of different authors. Part III deals with the application and consequences of given technologies in the context of developing nations in general and examines the issues of technology choices in development.

Michael Moravcsik explores the meaning of science, technology, and development and sets the standard for the assessment of science and technology in the context of developing countries. Of particular interest is Moravcsik's treatment of the value system, which according to him is the framework for science and the influence of science on our world view. The discussion includes his evaluation of the current national and international scientific cooperation and assistance efforts and how they can be assessed.

François Hetman strongly emphasizes the point that technology assessment is a systematic aid to better informed policy making, attempting "to discern ex ante possible patterns of development instead of just categorizing ex post experience." He notes that technology assessment as a discipline has come of age to recognize the "new questions, less of a methodological kind and more pertaining to its function, status within the policy machinery, use and usefulness in formulation of policy options." This means, as Hetman notes at the end of his chapter, having creative technology assessment to clarify available technological options and having creative policy makers to gear it in line with socially relevant issues.

Paolo Soleri's transcendent study targets the real issues in the design, introduction, operation, and maintenance of technology. Soleri would have us rethink some of the fundamental issues and notions about technology and what it has achieved for human evolution. Soleri's view of technology is that of a dynamic, organic, and transcendent organism that is constantly evolving to make the human being more perfect. He is a technological optimist, but he would like us to assess technology for its complexity, sensibility, and compassion rather than treat it as a complicated amalgam of large numbers of fitted parts. According to him, the mechanical association does not constitute technology, and neither should it be so assessed.

Mangalam Srinivasan's chapter includes a theoretical framework for assessing transfer and intended applications in solar energy technologies for rural India. Her conceptual approach to energy assessment recognizes the ordinate, substantive, and synergistic stages of assessments and includes the concept "sympodium," introduced by her to illustrate how issues and concerns are intertwined and superposed and how the analysis of one of the elements almost always involves the analysis of all other elements.

Viviane B. Márquez's study contends that technological assessment should include not only an evaluation of the impact of prospective industrial projects but also the analysis within a particular national context of patterns established by past technological choices that are believed to impose considerable limitations upon current and future choices aimed at reversing such patterns. She takes a critical view of the literature on technological dependency in Latin America. This is followed by a general description of the Mexican situation, which distinguishes the three levels of analysis, namely, national policy, organization, and craft. She discusses the possibilities of incorporating such findings in general principles of technological assessment designed to favor the emergence of technological development in national contexts similar to the Mexican case. Of special importance is her description of technological development, which is included as a special note to her essay.

Cheng Xuemin, who combines the roles of political adviser, technically trained engineer, and bureaucrat in the government of the People's Republic of China, is in an advantageous position not only to assess the hydropower development in his country (in which he has had an important role) but also to offer insights into decision paths as they relate to China's social and economic development goals. China has had good experience in hydropower planning in general. Most of its experience, however, has been in small hydro development, which has spectacularly benefited China's vast and dispersed rural constituencies.

This chapter is descriptive and offers valuable insights and lessons to those who wish to make use of the experiences. Political commitment and people's determination are elements that have caused enormous generating output to result in hydropower planning.

Momir Cecez reviews the matter of economic development in Yugoslavia through the use of technology. His chapter combines a historic view of technological developments as they occurred at various stages in the country's economic development and an outlook for the future. He addresses the various issues of political preferences, foreign technologies, technical training, and academic institutions in development. He concludes that technology, fixed assets, and employment are three factors that give significant impetus to the successful development of a nation and, therefore, should be included in any technology assessment effort.

Pilot-cosmonaut Valentina Tereshkova raises one of the most important issues in technological development—the role of women. Integration of women in technological progress and in the process of production, Tereshkova notes, is the measure of human development attained by a nation. Technology assessment can ensure such integration. Tereshkova reviews the Soviet space program in relation to women in the work force; their skills, education, and training; the special aspects of women in high technology; and how systems may be developed to help encourage greater participation by women. Herself an important member of both the policy apparatus and operational team, Tereshkova notes that most attempts at technology assessment turn out to be assessments of the technical content of a given project.

Avtar Atwal provides us with a descriptive account of the progress that has been achieved in agriculture in India in general and in his state of the Punjab in particular. The study of the Punjab agriculture is significant, because this state has shown the most rapid progress of any in the Indian continent in its subsequent increase in the standards of living for the state's citizenry. Atwal points out the most important aspects of this progress, including "assigning social status in rural India" through economic (agricultural) achievement and occupational status. He notes that "the functional utility of work is more valued now than the caste of a person." A distinguished educator himself, he is among those whose work both within the walls of the university and in the field have been responsible for the immense material progress of his state. Therefore, his evaluation, although descriptive, holds elements of assessment considerations. The technical analyses included in his chapter offer additional criteria for use in technology assessment in agriculture.

Michael Ansalem argues that studies of technology choices in the Third World have faced numerous decisions that have been outside the scope of neoclassical economic model explanations. Ansalem puts forward the theory that there are a number of factors influencing technology choice that do not change the relative price of the factors of production. These considerations, according to the author, have to do with the profile of the managers making the choice, with the management process that leads to the choice of a technology, and with management objectives and strategies that can be termed management variables in technology assessment and choice. His chapter examines a number of neoclassical assumptions and the different ways in which management processes and characteristics influence technology choice. Ansalem discusses the policy implications of the management factor in technology choice and how it may hold the key for achieving some of the most important economic development goals of developing countries.

In chapter 11 Donald Lamberton reviews some of the most important models of telecommunications experience in developing countries and concludes that "few technologies exemplify more fully the central problems of technology assessment than telecommunications." He notes that technology assessment of telecommunication developments offers the opportunity to make national policy choices that are capable of benefiting a large segment of society. New policy approaches made in Organization for Economic Cooperation and Development (OECD) countries and elsewhere may be advantageously studied for use in developing countries appropriate to each country's socioeconomic system. The emergence of "an economics of information," according to Lamberton, is to accept the "need to know a great deal more than is presumably known about the role of communication in development if technology assessment is to be applied to telecommunications development." This seems to be the message of the article.

Dilmus James offers a valuable and insightful analysis of the extent to which used machines are in demand in developing countries. James assesses the methodologically difficult matter of maintenance costs involved with used machinery. For example, he points out that in the case of some used machinery the maintenance costs are higher and their durability lower than new machines. On the other hand, the "obsolete" compressor imported into Mexico from Michigan is working to cool milk more rapidly and increase throughput. James points out that only a small part of the total used-machinery stock is sold to developing countries, while 95 percent is sold within the developed world. What about the problem of spare parts, maintenance, and the like? How can developing countries correctly assess the role of used machinery for their development planning? According to James, the following would be steps in the right direction: improved information administered by the United Nations; government nondiscriminatory trade policies with regard to the importation of used machinery; and national development planning, which takes serious note of the role of used machinery in industrial development. He supports these contentions by means of intense and significant research, relevant data, observations, and actual field investigations.

Maurice Middleberg and Frank Barros, taking the lead from the various discussions preceding their chapter, conclude that a joint management strategy is needed in effecting the transfer of relevant technologies responsive to the local development context. Reflecting upon the work of their own center, Middleberg and Barros propose that the essence of the joint management strategy between universities transferring technologies and the beneficiaries be one of shared power in decision making. This approach, according to the authors, opens up channels of information on needs, intersectoral dependencies, and user characteristics in addition to building a management capacity at the recipient level, which is so vital for technology to succeed and deliver the intended results.

## TECHNOLOGY ASSESSMENT AS ESSENTIAL KNOWLEDGE

Technology assessment, as a rational part of human behavior for evaluating tools and methods for prospective use, is not a newly invented discipline. The fact is that assessment of technology is a predisposition and is close to human survival, whether in the wild or in our currently mechanized, computerized world. Technology assessment cannot entirely eradicate risks but can ensure that a given technology comes within an acceptable level of risk, which, of course, inherently accepts the proposition that risk is an associated condition of human existence. Technology is just one way to human perfection in the constant flux of cosmic evolution, but it is also the most difficult and complex, carrying within itself the seeds of self-destruction. Great danger also lurks behind the fact that technology may be the final design to replace aesthetic appreciation of art and life and the poetry of living. Many creative artists and poets of this century and the last one have warned us that our worship of the furnaces will render our vision as "machines for making more machines."[4] Allen Ginsberg expresses his outrage at the fast-breeding building technologies and their molded plasticities and concrete boxes thus, in his "Howl":

> What sphinx of cement and aluminum
> bashed open their skulls
> and ate up their brains and imagination?
> Moloch, whose buildings are
> judgements![5]

Does technology assessment then have a role in enhancing our aesthetic and spiritual existence in addition to the material conditions of our lives? We think that it definitely has the responsibility. Technology assessment belongs to the essential knowledge of our individual and collective existence. It is knowledge that helps bring about change vital for our human evolution and knowledge that cautions us against arbitrary changes in work, societal, cultural, and spiritual values.

By changing what we know about the world, we are also able to change the world we know. Dobzhansky put it in the following manner: "Hope lies in the possibility that changes resulting from knowledge may also be directed by knowledge. Evolution . . . may conceivably be controlled by man, in accordance with his wisdom and his value."[6] That is the unifying theme of this book.

NOTES

1. Norbert Wiener, The Human Use of Human Beings: Cybernetics and Society (New York: Avon Books, 1967).

2. Robert Holt, "Technology Inducement Mechanisms and Technology Assessment," Working paper, Center for Comparative Studies in Technological Development and Social Change, University of Minnesota, p. 4.

3. Kierkegaard, The Sickness unto Death, quoted by Michael Washburn, "Outline for a Normative Forecasting/Planning Process," in Planning Alternative World Futures: Values, Methods, and Models, edited by Louis Rene Beres and Harry R. Targ (New York: Praeger, 1975), p. 60.

4. Gordon Bottomley, "To Iron Founders and Others," Poems of Today (London: Sidgwick and Jackson, 1920), pp. 46-47. Bottomley talks of those with mechanistic visions in the following manner:

You have brought down the fermament
And yet no heaven is more near
You shape huge deed without event,
And half made men believe and fear.

You worship in your furnaces
Which, like old idols, lost obscenes
Have molten bowels; your vision is
Machines for making more machines.

5. Allen Ginsburg, Howl and Other Poems (San Francisco: City Lights, 1971), pp. 9-22.

6. Theodasius Dobzhansky, Mankind Evolving (New Haven, Conn.: Yale University Press, 1962).

# PART I

## CONCEPTS AND TRENDS

# 1

# ASSESSMENT OF SCIENCE IN DEVELOPING COUNTRIES

MICHAEL J. MORAVCSIK

## TERMS AND OBJECTIVES

Assessment, as it is used in this volume, means the qualitative and quantitative rating of the object of the assessment in the framework of a certain value system. This chapter will deal with the assessment of science in developing countries. Our first task, therefore, is to specify what we mean by science and what the value system is in the framework for which we want to assign an index. We also have to analyze the aspects of science that we want to evaluate in this framework. Finally, we will have to articulate what we mean by the concept of "development." All of this will be done in the first part of this chapter.

The chapter continues by focusing on science in the developing countries in an attempt to see to what extent the concepts outlined in the first part can be and have been applied in that context. The chapter then turns to national and international scientific cooperation and assistance efforts and makes inquiries about the way in which and the extent to which such efforts can be and have been assessed. Finally, conclusions and recommendations are summarized.

### The Meaning of Science

Let us begin by specifying the meaning of science as used in this chapter.[1] Science is a human activity the objective of which is to learn about the world around us. We will restrict ourselves to the so-called natural sciences, which generate knowledge about the inanimate part of the world and about the physiological aspects of the animate world, including people. We exclude the so-called social sciences, which are interested in the collective behavior of animate beings and especially of people. The reason for the distinction between natural and social sciences is that at present the methodologies of the two are quite different; hence, the characteristics of the respective communities, the processes of progress, and the resultant methods of evaluation are also very different.

Note that the meaning of science is given in terms of its output: science is the activity producing knowledge about nature. Such a definition gives us an easy handle on differentiating science from technology: technology is a human activity the objective and output of which are prototype gadgets, procedures for doing things, processes, patents, and so on. This chapter deals only with science, since the rest of this volume gives ample attention to technology. There is, however, a connection between science and technology that will play a significant role in this discussion of the assessment of science.

Scientific research activity can be classified in several ways. One way is by the tools the researcher uses. Theoretical science describes that part of the activity in which empirical information about nature is used to discover a conceptual scheme in which natural phenomena appear ordered and explained in the sense of being reduced to the consequences of a relatively few general principles. In contrast, experimental science describes the process of gathering such empirical information about nature, including the design of experiments collecting such information. In some of the very advanced sciences, like physics and chemistry, these two activities are quite sharply differentiated in that a given person is usually trained for and engaged in only one of the two. In the less well-developed sciences, such as biology, such a sharp distinction is not yet present (mainly because the theory of biology is still extremely rudimentary), and the same researcher is generally engaged both in experimental work and in whatever theory may exist.

Scientific activity can also be classified by the motivation of the researcher or of the entity supplying the support for scientific research. According to this, scientific research is divided into basic research and applied research. The motivation for the former is primarily intrinsic; that is, the justification is in terms of enlarging scientific knowledge itself. The motivation for applied research is primarily external to science; that is, the desire is to utilize the knowledge for technology or for other nonscientific tasks. The distinction between basic and applied science (in contrast to the distinction between theoretical and experimental science) is quite tenuous and ambiguous. A given piece of research usually has both basic and applied motivations behind it. Furthermore, it may be basic to one person and applied to another person. In particular, it may be basic to the researcher himself and applied to the organization supplying the resources for research. It might also be basic in the short term and applied in the long term. In practical science policy, nevertheless, it is useful to attempt to make at least an approximate distinction between these two types of research. Thus, one might say that a piece of research is applied if the primary motivation for performing it on the part of the researcher himself is the expectation that the results will be used in the next ten years or so for a purpose external to the body of scientific knowledge itself. It is misleading to say, however, that a piece of research that does not find such an application must therefore be basic, because this would also include poor-quality applied scientific research, which never finds an application because of its inferior quality or because of the lack of a mechanism to connect it with technological development work. This situation appears quite often in developing countries.

It goes without saying that applied scientific research and technological developmental activity are two quite different activities, though they may be related in the sense that the latter may depend on the former. They may also be connected by the formation of a team in which both applied research scientists and developmental engineers are included. It is, nevertheless, important to keep in mind the distinction lest one nurture false expectations and make inappropriate arrangements for the staging of these activities.

It also goes without saying that the dichotomy of theoretical versus experimental science and the dichotomy of basic versus applied science are totally different. Accordingly, there can be four different types of research activities: basic theoretical, basic experimental, applied theoretical, and applied experimental. The confusion of the two dichotomies and the resulting errors in science policy decisions are also frequent in developing countries.

## The Characteristics of Science

Science (basic or applied) is characterized by its methodology. The search for empirical knowledge is through experiments and the ordering and explanation of the observations into theoretical concepts, which are then confirmed (or, more precisely, falsified or not falsified) by further experiments that these theories suggest. This methodology has brought forth four important characteristics the natural sciences have had for the past 400 or so years: objectivity, collectivity, universality, and cumulativeness.

Objectivity means that the community of scientists[2] has a set of criteria to tell who is right and who is wrong, which works in the overwhelming number of cases and overrides any nonscientific distinctions that may exist among scientists, such as age, nationality, religion, sex, race, ideology, or personality traits. Collectivity means that scientists depend on each other in their scientific work to a very large extent, and that a piece of research makes a mark in the history of science only if it is widely dispersed within the scientific community. Universality means the already-mentioned independence of science from nonscientific parameters and the ease and ubiquity of scientific collaboration across all such nonscientific boundaries. Finally, cumulativeness means that progress in science follows an arrow, that there is a definite and unambiguous way of assessing progress in science; and so one can say with confidence that a staggering amount of progress has been made in science since, say, 1650. A similar statement cannot be made with any conviction and unambiguity in any other human activity[3] (such as the arts, humanities, politics, social sciences, philosophy). In these other realms we have also added to the previous stock of accomplishments, but there is no generally agreed-upon way of ascertaining that the new addition is in any way better than what came before. This cumulativeness of science (basic or applied) is exactly the trait that allows us to talk about assessment in a fairly meaningful way.

All the above considerations about science are subject to some exceptions (the existence of borderline cases, the imperfections of the methodology in the short-term sense, and so on). There is also the apparently beclouding issue that paradigm changes[4] in science can produce conceptually radically new ways of looking at nature that seemingly destroy the previous theories and, hence, apparently negate the claim for cumulativeness. One must remember, however, that even if the conceptual aspect of scientific theory appears sometimes to be noncumulative, the predictive power of science is cumulative even across seemingly radical paradigm changes. In any case, it is safe to claim that in comparison with other human activities, science does have the above characteristics, and that, for the purposes of science policy decision making, these characteristics are very important guiding principles.

## Is There More than One Science?

There is only one science. This follows from the above considerations.[5] This one science is not Western science, though it happens to have been developed primarily in the domain of the Western civilization during the past 400 years.[6] In this century a number of other parts of the world having cultural heritages radically different from Western civilization have entered science making and have become excellent in it. Perhaps Japan is the best example. The science these countries have evolved is, however, indistinguishable from the science generated by the Western countries, since there is only one science.

This point may appear to be an unnecessary one to stress except for the fact that there are a number of groups, particularly in some developing countries, who have claimed that there are or could be several different sciences, and that therefore their own country should not have Western science foisted on it.[7] It is important to mention this in the context of this volume, since, if the claim is valid, the assessment of science in developing countries would have to be done in a possibly significantly different way from the method of assessment used, say, in the United States.

To the best of my knowledge, however, to date such claims of the existence of several sciences have never passed beyond the stage of vague talk. I know of no example suggested by anyone for such a different science, and I do not even know any description of how such a different science might look. The example of African witch doctor science is in no way to the point, since empirical phenomenology without a theoretical framework is in fact part of the science and has been in existence throughout the history of science. Some proponents of the many-science hypothesis give examples in which the use of scientific knowledge (or, more precisely, of the technological developments based on such scientific knowledge) are different from those generally found in Western societies.[8] This, of course, is not in dispute: the scientific method says nothing about the way such technological developments are to be used; hence, we have complete freedom, from a scientific standpoint, to formulate uses that may indeed be different from culture to culture and from situation to situation. In that sense the one- versus many-science controversy is sometimes a purely semantic one. Beyond that, however, one must repeat that there has been evidence brought forward for the existence of more than one science. This will be kept in mind throughout this chapter in the discussion of methods of assessment of science in the developing countries.

## The Value System

Under what value system is science to be assessed? In other words, one must discuss the aims, objectives, and/or goals of science. In doing so, one apparently runs into a hopelessly chaotic situation. If we ask about this among different people in different countries with different political, religious, ideological, philosophical, and professional outlooks, we get at least as many different and even contradictory answers. After some thought, however, it is possible to sort all those answers into three main headings,[9] which thus can constitute the goals or objectives of science. These are science is the basis for technology, science is a human aspiration, and science has an influence on a person's view of the universe and of one's own role in it.

That much of the important technology is now science based is generally recognized. Up to about 100 years ago non-science-based, trial-and-error-based technology played a dominant role; even today some of technology is trial and error based, at least in its immediate, direct antecedent. But by now this is the minority of technology. Virtually all technological developments today utilize a substantial amount of scientific knowledge:[10] some generated a number of decades previously, some brand new. Inasmuch as technology is essential in satisfying both the material and nonmaterial aspirations of people, science as a basis of technology is ranked highly. Of the three broad motivations for science, the technological one is the most obvious one; hence, it is not necessary to dwell on this at great length.

The second broad area of motivation for science is based on the claim that science is an important human aspiration in the twentieth century. Human aspirations—material and, particularly, nonmaterial—govern human behavior and determine the fate of individuals, nations, civilizations, and humankind as a whole. Curiosity; sense of achievement; power; prestige; the propagation of religious or political principles and systems; aesthetic satisfaction; philosophical convictions; the service of God; the sense of national, cultural, or racial identities; and many other nonmaterial aspirations have played crucial roles in human history and continue to do so today, as countless examples from the present-day world scene illustrate. Individuals or groups of individuals without such nonmaterial motivations falter, decline, lose their energy and élan.[11] In fact, most individuals or groups of individuals define the purpose of life and the purpose of their existence in terms of such nonmaterial goals. Contemporary psychology, even though still far from being a respectable science in the sense that word is used in this chapter, increasingly bases its analysis and methodology on such nonmaterial elements in the human psyche.[12]

The pursuit of scientific research and excellence in such activities are among the important nonmaterial aspirations in the twentieth century. They contribute to a number of the aspects listed above, including aesthetic feelings, acquisition of prestige and power, the reinforcement of national identities, sense of achievement, curiosity, and others.

This point must be especially emphasized in the context of developing countries. The state of being a "developing" country is bothersome; it is a burden; it is debilitating not only and not even primarily because of the sizable difference in the standard of material living (the GNP/capita). Much more acute than pure material poverty is the total dependence of the developing countries on the "developed" countries in every walk of life (economic, cultural, scientific, political, artistic, social, and so on). Intentionally or unintentionally, the countries with the know-how and knowledge dictate the pace of the world; the other countries have little choice but to participate in this or isolate themselves into stagnation. The ubiquitously emphasized dependencia found all over Latin America is one well-known articulation of this state of affairs, but the feelings and the behavior of other countries and other parts of the world echo the same.

Individuals and groups of individuals expect and hope to realize their own aspirations and are immensely frustrated if they are relegated to the role of a follower and are prevented by the state of affairs to determine their own fate. Thus, having its own scientific infrastructure and participating in scientific activities on an equal footing with other parts of the world are among the aspirations of developing countries and of individuals in those developing countries. This is fully reflected in many aspects of the organization of science in those countries, including prestigious institutions, a preoccupation with external recognition, a love for the symbolic presence of complex research instruments, and so on. When touring the scientific establishment of developing countries and interacting with science policy decision makers in the developing countries, it is difficult to avoid the conclusion that this aspirational dimension of science plays an extraordinarily important role there. This is quite legitimate and cannot (and should not) be dismissed as nonfunctional.

The third broad domain of motivation for science is in terms of its influence on our world view. In Western civilization, on which science as a cultural and philosophical force has exercised its influence for some 400 years, it is difficult now to appreciate the enormous impact of this influence on our way of thinking and on our attitudes. I do not mean here the question of how many people in the population as a whole can recite Newton's laws or the second law of thermodynamics. What matters much more are the attitudinal and methodological elements of science: its appeal to empiricism as an arbitrator, its optimism in terms of the existence of answers to questions, its belief in the openness of knowledge, its implication for a people's ability to influence their own fate, and so on. Science, while not an entire culture, is a most powerful cultural element that can be fitted into many different cultural settings to bring about a considerable change in the attitudes and behavior of the members of such cultures. The desirability of change is itself such an influence of science.

While in the developing countries the influence of technology has in many cases been considerable, the influence of science as a cultural and attitudinal force is only at its beginning, since it began only a few decades ago instead of centuries ago, as in the case of the West. If the huge gaps between the developed and the developing countries are to be narrowed; if the dependence mentioned earlier is to be lessened; if the world is to be directed toward a state in which its parts are on a more equal footing; if the various segments of the human population are to benefit comparably from new and impending developments (for example, space colonies[13]), we must allow all parts of the world to benefit from the cultural impact of science. Components of development have many of the same characteristics as science,[14] even though by no means is all development science related. Thus, science in the developing countries can play a pivotal role in this attitudinal sense.

Each of these three broad areas of motivation for science contain scores of special implications and applications, which may be taken into account when a specific scientific project is assessed. What was outlined above should, therefore, be taken as a general framework into which the various specific considerations and justifications can be placed.

## Activity, Productivity, and Progress

Having now defined the general value system in which science will be assessed, it is necessary to analyze the aspects of scientific activity that need to be examined when making an assessment. In particular, it is important to make a distinction among three such aspects:[15] activity, productivity, and progress. The distinctions among these three concepts might best be described by an example of what one would see and look for when visiting a scientific institution. Activity would be represented by experimentalists building equipment and taking measurements on them, theorists writing formulas on pieces of paper, others placing inputs into computers, science teachers lecturing to classes, and so on. Productivity could be discerned by looking at the results of the experiments, the outcome of the theoretical calculations and computer programs, or the effect of the science classes on the students. Finally, progress could be gauged in terms of the extent to which a scientific problem was actually solved by the previously mentioned activity and productivity or by the extent to which an outstanding scientist is created by the educational activities and productivities.

Activity is the most conspicuous of the three and also the easiest to measure by a person untrained in the sciences. It is also the least important of the three in that it is only indirectly, if at all, related to the three main objectives of science listed earlier.

Productivity is more difficult to measure, since it requires a scientifically trained person to assess the results, to test the caliber of the science students, and so on. Productivity is, however, more important than mere activity in terms of our objectives.

Finally, progress is extremely difficult to measure even for a scientifically trained person, especially if the measurement has to be done in a contemporaneous way. It is easier to evaluate progress made in science a hundred years ago when we have the hindsight and perspective to see the results in a broader context. But a historical assessment of science is relatively rarely the task at hand: science policy has to deal with the assessment of the very recent past, or the present, or even the future. Progress in that time frame is very difficult to ascertain. In the ultimate analysis, progress is exactly what we want to measure: it will tell us the extent to which the aspirations embodied in the three broad motivations laid down for science have been realized.

The difference between activity, productivity, and progress can also be illustrated with an analogy. If we want to get from a dense forest to a gold mine, activity would consist of the amount of thrashing around, the amount of cutting tree limbs, the amount of underbrush cleared. Productivity would be measured in terms of the number of feet of trail generated by the activity. Finally, progress would be measured by the extent to which our distance from the gold mine was decreased by the trail making. We do not definitely know this last measure until we have, in fact, found the gold mine.

## Quality, Importance, and Impact

A somewhat different trichotomous distinction in the assessment of scientific efforts can be made along the lines of quality, importance, and impact.[16] Quality pertains to a description of how well that scientific effort was done. In the case of research, it would discuss the extent to which the work is free of errors, the degree of aesthetic beauty of the research, the amount of originality in it, and so on. Importance is gauged in terms of the surrounding research activity and describes the potential influence of that piece of research on subsequent research and on the solution of that and similar scientific problems. Finally, impact describes the actual influence of that piece of research on the surrounding research activities. This measure, therefore, depends on, in addition to quality and importance, the degree to which the communication system of science was used successfully in this instance to propagate the results of the research within the scientific community and to incorporate them into the universal body of scientific knowledge or into the knowledge system utilized by technological development work.

This trichotomy is also very important, especially in the context of developing countries, since it views scientific efforts not only in the purely abstract, cognitive way but also in terms of the processes within the scientific and technological community that link the cognitive with the operational and collective factors in science.

## Development

All that remains now for us is to agree on what we mean by development.[17] Development in this discussion is the course of action taken by an individual or a group of individuals in order to achieve a greater realization of their aspirations. The central role aspirations play in the life of individuals and societies was discussed earlier; hence, the concept of development should be tied to such aspirations. The concept, therefore, involves efforts in the direction of both material and nonmaterial aspirations. This is an important point to stress, since the way development is often used makes it synonymous with "economic development" or "material development." Not only is that an unreasonably and unrealistically narrow definition of development but it is also a nonfunctional one; often economic development problems cannot be meaningfully separated from development problems of other sorts: psychological, cultural, social, political, attitudinal, religious, and so on. In tune with what was said earlier, the term developing countries designates those countries in which the process of development in the past was severely hampered, but which presumably are now in the state of making efforts to accelerate and broaden that process. In contrast, developed countries are those that presumably have a greater opportunity to realize their aspirations and have gotten farther in doing so. The terminology is a very unfortunate one, compounded by the growing existence of a third category of "undeveloping" countries (in some respects the United States), namely, countries in which the opportunities for realizing aspirations are decreased partly by ideological design and partly by ideological and philosophical blindness. Be this as it may, however, this chapter will use the word development in the above-defined broad meaning and assess the developing countries in their scientific activities by considering values stemming from a wide variety of aspirations, not only material or economic ones.

This concludes the discussion of terms and objectives. The reader may have grown impatient by now, since a significant fraction of this chapter has been taken up by seemingly preliminary tangentialities. This accent on the definition of the framework was intentional, however. Assessment is, by its nature, a complex problem that relates to a broad assortment of considerations; hence, breadth and rigor in its analysis are essential if we are to make an assessment that has a functional impact on science in the developing countries and is not merely a frosting on the cake designed for committee reports, agency filing cabinets, or ceremonial development plans.

## ASSESSMENT OF SCIENCE

This part of the chapter will discuss the status of the actual assessment of scientific efforts in developing countries and the related methodological problems involved in such assessments. Scientific efforts, just like any other human activity, can be visualized as an input-output process. Certain resources (in the case of science they are manpower, funds, equipment, materials, buildings, and so on) are fed into a "black box," and certain products then emerge from the box as a result of the input. It is granted that this way of looking at science development is somewhat simple-minded, modeled after the economist's picture of a factory manufacturing cans of tomatoes; it jumps over the most crucial and most interesting part of the problem, namely, what goes on inside the black box that transforms the input into the output. Yet, from the point of view of the net effect of scientific efforts and of the impact of science on the country and the world, ultimately it is this relationship between input and output that counts.

### Overemphasis on Input

The first important observation one can make when looking at efforts to describe this input-output process for science in developing countries is that an inordinately large (in fact, overwhelmingly huge) fraction of the data, analyses, plans, and conclusions deals with the input rather than with the output. This is evident in national development plans; national papers written for the United Nations Conference on Science and Technology for Development (UNCSTD); country surveys by international scientific assistance organizations (like those of the Organization for Economic Cooperation and Development [OECD][18] or the United Nations Educational, Scientific and Cultural Organization [UNESCO][19] or the Organization of American States [OAS][20]); and country surveys by national scientific assistance agencies (like the U.S. Agency for International Development [AID][21]).

The reason for this state of affairs is easy to comprehend. The measurement of input, at least on a rudimentary level (that is, ignoring the quality factor), is a relatively unsophisticated task of collecting statistics on mundane items like money, size of personnel, numbers of organizations, and so on, a task that requires little if any scientific training. In contrast, the assessment of the output of science is a highly complex process for which methodology exists only in a fragmentary way and which, in any case, requires a considerable amount of training in the sciences, as well as the cooperation of scientists themselves.

There is also another reason for the assessment of input being so prominent compared with that or output. As was mentioned earlier, one of the three main motivations for scientific ability is the participation in a strong human aspiration. From this point of view (that is, from the standpoint of demonstrating that a country is in fact "doing science") the exhibition of input is, for many purposes, just as satisfactory as the gauging of the output. Reports listing the thousands of scientific personnel, edifices bearing the proud inscriptions denoting them as halls for research, and columns of expenditures earmarked for scientific activities in many respects are ipso facto demonstrations of the country's resolve for a capability of conducting scientific research. To be sure, the knowledgeable worldwide scientific community itself will not be impressed by the exhibition of input factors by themselves, because to them only the output counts; but this community is a tiny and politically insignificant one compared with the overall audience that watches a country's image.

There are a number of detrimental effects resulting from this virtually exclusive preoccupation with input. First, viewing input as divorced from output leads to extensive central planning operations for science, producing organization charts, national science councils and committees, research priority listings, and other formal undertakings, the futility of which cannot be ascertained in the absence of gauging the output. In fact, the impression is often created that having prepared such charts, plans, and listings, 99 percent of the job is done, and the country has thereby "acquired science." In actuality, such operations, even if done appropriately, constitute no more than 5 percent of the job to be done.

Second, an obsession with input obscures the real goals of science as previously outlined. It is possible to make all the arrangements for input without having to consider in a serious way and in depth the objectives and goals of science. The point is that regardless of what these goals and objectives are, science needs people, money, equipment, and so on. To be sure, the analysis of the objectives helps in deciding what kind of people, equipment, or buildings are needed and for what the money should be spent. But such details are seldom considered anyway, at least on the planning level. In fact, quality is rarely taken into account when specifying input. When manpower targets are given, nothing is mentioned about the quality of the scientists-to-be and about how the desired quality can be obtained. Thus, not only is the preoccupation exclusively with input, but much of the time is spent only with the quantitative input. Using the terminology previously developed, one might say that such input is related at best to output activity, but not to output productivity or progress.

## Research Output

So more attention is needed to the assessment of output. This is, however, easier said than done. Let us review briefly the various functions of science and the ways in which the output can be measured in those functions. From the points of view of both the first and the second motivations for science (that is, science as a basis for technology and science as an aspiration), the most important function of science is research. Therefore, it is necessary to review in a nutshell the existing methods of assessing the output of scientific research.[22]

All such methods of assessment are based on the assumption that a piece of scientific research is valuable only if it is communicated from the researcher to the scientific community at large; hence, one can use the communication system of scientists for measuring research output. This is a reasonable assumption in that a piece of research, the results of which die with the researcher without having been communicated, fails to be incorporated into the body of scientific knowledge and fails to be utilized by technologists.

After this assumption is made, the assessment methods of science divide into two types: the assessment through a monitoring of the scientific literature and the assessment through peer review, that is, through the personal opinions of scientists. Both methods have their advantages and drawbacks and, therefore, should be used together. They are not quite independent from each other because the scientific literature is written by the same scientists who give their opinions. Yet, the two methods are not completely overlapping either. Peer assessment is more capable of taking into account those contributions of scientists or groups of scientists that leave no direct trace in the written literature, such as informal discussions and communications, conference contributions, and participation in policy formation and problem choice. On the other hand, peer review generally is done with "small statistics" utilizing only a few or at most a few hundred scientists, while the monitoring of the literature can nowadays be done with the use of computers and, hence, on a scale of millions of articles and citations.

It should be added that peer review evaluation is extremely rarely done in developing countries. The domestic scientific manpower is usually too small and often too skirmish ridden to be useful for the purpose, and there is an enormous amount of reluctance on the part of developing countries to use an international group of peers in the assessment of science in the country. This is partly owing to a feeling that the use of such a group is tantamount to the kind of dependence that was previously discussed and is unacceptable. Inasmuch as the scientific community is worldwide, and even the scientifically most advanced countries use international groups for assessing their science, this feeling on the part of the developing countries is entirely misplaced. There are also other factors contributing to the absence of such international peer review in the developing countries. In order to select, contact, and engage such a peer review team, a considerable amount of scientific expertise must already be present in the developing country and, in particular, in the "science policy" agency that is (or should be) concerned with such an assessment. That is rarely the case, however.

## Scientometric Indicators

The methods used for assessment through the monitoring of scientific literature are generally based on the journal literature, since that is the most accessible to computer handling. The actual quantities used range from the number of scientific authors (that is, the number of scientists who published at least one article in a scientific journal in that year) to the number of articles published and to the number of citations received (that is, the number of times an author or a group of authors was referred to in other scientific articles). Generally, the more sophisticated such a measure is, the more appropriate it is to be applied to a small set of scientists and the better it incorporates quality in addition to quantity. For example, the number of scientific authors might be useful in comparing the scientific output of the two countries, but is generally useless when comparing two scientists competing for the same position. Also, citation counts are more indicative of the impact of a scientist than publication counts, since presumably a higher citation count indicates that the original article was more useful for subsequent research. This assumption is not quite justified, because an article can be cited in another article for a variety of reasons besides having contributed substantially to the citing article.

Such scientometric indicators of the scientific output in developing countries are beginning to have a considerable literature.[23] Information is beginning to be available on the number of scientific authors in absolute terms and per capita, the distribution of the various scientific disciplines in different countries or geographical regions, and so on. Curiously, but perhaps not unexpectedly, a large segment of the authors producing such information about developing countries is not from the developing countries themselves, though some exceptions exist.[24]

Attempts have been made to correlate the scientific "size" of a country, measured by such scientometric indicators, with some other nonscientific parameter characterizing the country. Perhaps the most successful of these has been the correlation of the number of scientific authors in a country with the gross national product (GNP).[25] Both of these parameters vary from country to country by many orders of magnitude, but the correlation holds in most cases within an order of magnitude. The accuracy of the correlation can be improved[26] if the degree of per capita affluence of the country is also taken into account by dividing countries into "rich" and "poor." In that case each kind of country correlates even more accurately, but the two correlation factors differ from each other somewhat. That such correlations should exist can be "derived"[27] in an approximate way by considering the cost of research per researcher, the percentage GNP spent on science by various countries, and the efficiency of the science dollar in various countries. It should be remembered, however, that the correlation is only approximate, and that within it there is much room for the difference between a highly motivated, high-quality scientific establishment supported by judiciously spent funds and a listless, mediocre science infrastructure burdened with a heavy and incompetent bureaucracy.

A very interesting and very important facet of scientific productivity is the distribution of individual productivity within a scientific community. Using data for the scientifically advanced countries, the so-called Lotka's Law[28] has been established according to which the number of scientists producing $\underline{n}$ research papers in their lifetime is proportional to $1/\underline{n}^2$. This means that the productivity of scientists is highly skewed: for every 10,000 scientists in a scientific community who publish one research paper during their entire career, there are only 100 who publish ten, and only one who publishes 100. Various conclusions can be drawn from this fact with regard to science policy decisions.

As mentioned, Lotka's Law was established on the basis of data from scientifically advanced countries. No corresponding information is available about the scientific communities in developing countries. It would be most interesting to have such data, because it would help resolve a controversy (which in itself has wide-ranging science policy implications) about whether Lotka's Law is a direct result of the internal rules of the "science game" alone or whether the law can be significantly influenced by the environment, in a broad sense, in which the scientist operates.

The scientometric indicators usually used are based on data collections in the large international centers of scientific information, such as the Institute of Scientific Information (ISI) in Philadelphia. Whereas these collections are based on surveys of most of the scientific articles published around the world, they encompass a relatively small fraction of all the scientific journals published worldwide. Typically left out are the small and infrequently appearing journals published in developing countries. As a result, it might be claimed that authors in the developing countries are underrepresented in the data collection, since they are more likely to publish at least some of their articles in the omitted journals than authors in scientifically advanced countries. There is some truth to this claim, though it is not likely to affect seriously the comparison between two developing countries or the comparisons made within a developing country.

There is a relatively easy remedy to this deficiency. Since most authors in developing countries publish either in the well-known international journals or in local journals of their own country (and, in a few cases, in regional journals), it would be easy to produce a more complete productivity count for authors in developing countries if each country established its own data bank about the authors, articles, and citations in all the journals published in that country. The smaller the country and, hence, the more primitive the facilities may be to collect and handle such data, the smaller the job would be; therefore, the scheme is quite feasible. To catch the few regional journals, they could be added to the information system of each country in the region. This network of information data banks would be much preferable to another scheme, also proposed recently, in which some central organization would produce a data bank for all research published in developing countries. To do the job in this latter way would involve the scanning of tens of thousands of journals; one would also have to overcome considerable obstacles in distributing the resulting information from the one central place to all those in the world who want to know it. Establishing indigenous scientific data banks in each country would also have some educational benefits for the individual countries in the areas of science management and data handling.

It is sometimes charged that while scientometric methods are appropriate for the evaluation of the results of basic scientific research, they fail for applied scientific research because the results of the latter tend not to appear in journals, especially when the research pertains to a local applied problem. This contention is unwarranted and is based on a confusion between applied scientific research and technological development work, a confusion that is common in science policy circles, particularly in developing countries. Indeed, the results of a technological development project often do not appear in journals, since the results are in the form of a prototype, process, patent, or procedure for doing something. In contrast, the results of a piece of applied scientific research, if of good quality as judged by its own internal criteria, should always appear in a scientific journal, local or international. The product of such research is knowledge and not a tangible piece of hardware, and the only way to communicate it to others and to preserve it for later users is to publish it. This can be done in the form of a report, but that is vastly less effective from the point of view of communication than depositing it in a journal, even if the journal is small and local.

## Assessment of Impact

The above peer review and scientometric indicators assess only scientific research itself and not the extent to which it contributes to the objectives of science. One might say (using the terminology previously developed) that those indicators measure quality and perhaps importance but not impact. So how can such impact be assessed?

From the point of view of the first type of objective of science (namely, the contact with technology), the transfer must be made through engineers and other technologists who are to utilize the scientific knowledge in their development work. This link in the science-technology-production chain is usually very weak in developing countries.[29] At the same time there is no methodology or mechanism to measure it either. The process of transferring knowledge from the scientist to the technologist is often in the form of personal contact or in the form of the latter reading the article or report of the former; neither of these events leaves a permanent trace in the record (unlike, for example, the informal communication between scientists, which sometimes leaves the trace of a citation or acknowledgment). The best way to assess this link would be by a detailed investigation of the origin of instances of successful technological development work in developing countries. This has rarely been done.

In this connection another, very important point must be made. The share of an average-sized developing country in the world's scientific output is on the order of 0.01 percent to 0.1 percent. In other words, out of 10,000 scientific papers written in a given year, only one to ten are likely to have been written in the country itself and an even tinier fraction of them by one given scientist. When, therefore, a scientist interacts with a technologist in order to help him with scientific information in connection with a specific technological problem, the relevant knowledge will, in the overwhelming fraction of the cases, not come from this scientist's own personal research but from the general body of scientific knowledge in the field of that scientist, most of which was generated by others and abroad. This is even true in the case of applied research, when the knowledge required pertains to local parameters. Thus, from the point of view of the direct generation of knowledge, the personal research work of the scientist is of small importance compared with his overall knowledge in his field.

The crucial point, however, is that the only known way to maintain a scientist knowledgeable about his field is to support him in his personal research work. In this sense personal research is an overhead[30] expense paid in order to assure scientific competence in a much broader area. As a result, the exact topic of personal research is of little importance. This is a point seldom recognized in developing countries, where elaborate arrangements are often attempted to make sure that the personal research of the scientist be relevant.

It might be claimed that the transfer process between scientist and technologist can be assessed by looking at the end product in the chain, such as patents,[31] prototypes, processes resulting from the development work, or even by the output of the productive sector based on that technology. This is a dubious contention, however, since a failure in the end product can be owing to a number of defects in the chain, many of which may have nothing to do with science or the scientist. For example, if the country's engineering manpower is of low quality and poorly educated, as may often be the case, no amount of brilliant, applied scientific research or group of talented, communicative, and well-informed applied scientists can produce a result. The remedy, however, is not to try to make applied scientific research more relevant but to take measures to improve the stock of engineers, something that requires science policy measures that may be completely different from those aiming at applied scientific research. It is, therefore, necessary to investigate specifically that part of the transfer process that takes place between the applied scientist and the technologist.

## The Other Roles of Science

The assessment of scientific research in the context of the other two objectives of science (namely, to fill an aspiration and to affect one's view of the world) is also necessary. To a large extent, both of these tasks pertain to people outside the scientific community. Therefore, one has to somehow assess the efforts in science education and the efforts in mass communication with the bulk of the population. Here there is a lack of methodology and a lack of tradition in evaluation. Educational systems tend to be ingrown and produce their own evaluative systems by setting up an examination system plagued by the same defects as the education itself. In the case of mass communication through newspapers, radio, exhibits, and possibly television, the task is even more complex. In this latter area much remains to be done in developing countries. It would be especially important to communicate to the population as a whole the attitudes, philosophy, methodology, and value system of science, not only the facts unearthed by science. To assess such an activity, it would not be enough to list the number of hours science programs are beamed over the radio waves or the amount of money spent on the local science museum. These are input quantities, and one is mainly interested in output. Methods of social science, if applied knowledgeably, would be of great value in assessing progress in mass communication about science.

Finally, we might also want to assess the administration and management of science in a given country. Here again, well-tried methods are few, and precedents are rare.[32] It has seriously been suggested that a measure for this should be the number of science councils established.[33] Not only is this again a matter of resorting to input rather than measuring output, but even on the input level it sounds preposterous to many seasoned observers who have seen the damage that such formal bodies can do to the local scientific community. In some cases, of course, such bodies—or, more precisely, outstanding individuals[34] under the umbrella of such bodies—can play a crucial role in the formation and maintenance of the local scientific infrastructure; but this is a sufficiently chance phenomenon to preclude a reliance on the mere number of such bodies as a policy indicator. Here also, detailed case studies can produce much more valuable information than can statistical methods. It would therefore be very important to have objective and accurate accounts of the work of many outstanding science administrators in a variety of developing countries. Since such exceptional people almost invariably earn the envy and hatred of their more mediocre colleagues (sometimes justly so, since such pioneers often are autocratic and too singleminded about their goals), such accounts would have to be written by knowledgeable outsiders who are difficult to find.

## ASSESSMENT OF INTERNATIONAL SCIENTIFIC COOPERATION

As mentioned earlier, science is a collective and universal activity, and the international dimension of science is a central one. Furthermore, since modern science was first developed in the West and developing countries are still in the beginning of their participation in science making, for them international assistance and cooperation are crucial undertakings. Therefore, a brief discussion of how such international cooperation can be assessed is appropriate.

To begin with the conclusion, there is at the present no significant and functional methodology for doing such an evaluation, and partly as a result, no such evaluation has been done. Therefore, one should discuss what organizations need such assessment, why there is no methodology for doing it, and what can be done to formulate one and thus get the present stagnant situation off dead center.

## International Organizations

There are, roughly speaking, three types of organizations involved in international scientific assistance and cooperation. The first type consists of international organizations like the UN special agencies, regional bodies like the OAS, intergovernmental agencies like the OECD, and so on. They are characterized, in general, by being tied to governmental channels, having high visibility, and, hence, being very strictly constrained in their programs, both in substance and in appearance. They are usually very conscious of the importance of "not rocking the boat" lest their political balance (often precarious) be upset. They also must communicate in developing countries with the appropriate branch of the local government rather than with the scientific community itself; since these two are often sharply separated and isolated from each other, the input of the local scientists into the determination of programs and into their assessment may be very tenuous.

It is not difficult to see that in this interaction between these international organizations and the local governmental bureaucracies, the existing natural forces do not act in the direction of a functional evaluation of programs by criteria that would be meaningful to the local scientists. This does not mean that good programs do not occasionally occur,[35] since in any system there are individuals who manage to break through the shackles of the system itself and create something based on a right vision and resulting in real benefits, as judged by working scientists themselves. In such cases some assessment does take place, but its traces are very subdued, very anecdotal, and obtainable only by personal acquaintance with a significant number of scientists who are involved. To the best of my knowledge, no international agency has ever made an attempt to assess its work through a large-scale, systematic survey of the opinions and critiques of the scientific communities in developing countries themselves.

To be impartial to all sides, it should be added, however, that the blame cannot be placed entirely on the organizations themselves. The fact is that the scientific communities in developing countries themselves are also to blame, since they are remarkably passive and silent about their own problems and about constructive suggestions and proposals for improvements in international scientific cooperation. There are, of course, always some exceptions, but on the whole the overwhelming part of the existing literature on problems of science in developing countries originates with people in developed countries.[36] There is no reason whatever why a scientific society in a given developing country could not organize its own project to assess the effectiveness of, say, UN projects in that country. Yet, no instance of this has happened.

To be sure, as was said earlier, no overall and operational methodology has been articulated for such an assessment. It has been suggested to at least one UN organization[37] that it arrive at such a methodology by bringing together 20 people knowledgeable and experienced in practical matters of science management and assessment, secluding them on a "desert island" for two weeks, and giving them no other mandate than to produce, at the end of the two weeks, a functional assessment methodology to evaluate the work of an international scientific assistance and collaboration agency. The method need not be a theoretically optimal one, and, in fact, it could not be, since many theoretical problems connected with such a methodology remain unsolved. But the method could represent the best in the "state of the art" in this field and should be practically applicable. I have no doubt that such an exercise would result in a procedure that could be applied and could generate interesting results. The project of formulating such a methodology would cost something on the order of $100,000, a paltry sum when compared with the hundreds of millions of dollars international agencies spend on scientific cooperation and the comparable or larger sums spent by national agencies that could also utilize the results of the project.

## National Organizations

The second type of organization involved in international scientific assistance and cooperation is the national governmental agency. Some examples are AID or the Division of International Programs (INT) of the National Science Foundation (NSF) in the United States, the International Development Research Centre (IDRC) and the Canadian International Development Agency (CIDA) in Canada, or the Swedish International Development Authority (SIDA) and the Swedish Agency for Research Cooperation with Developing Countries (SAREC) in Sweden. One might also add to this list organizations like the Board on Science and Technology for International Development (BOSTID) of the U.S. National Academy of Sciences (NAS), which although officially only a semigovernmental body, depends for its existence almost entirely on governmental funds. Although this type of organization is also plagued by the difficulties mentioned earlier in connection with international organizations, the problems in some of them are less severe, and there are opportunities for more flexibility and a more direct interaction with scientific communities. Correspondingly, there are occasionally healthier signs in evidence in such organizations with respect to an attempt to assess their own activities.[38] In the instances on record the methodology used was ad hoc and was constructed by the organization to be assessed. While it is true that insiders know more about some of the details of the operations and, hence, can ask questions pertinent to those, it is also true that the fresh approach of outsiders might add breadth and scope to a possibly fossilescent organization. Hence, the optimal and yet still ad hoc method of constructing a methodology would be through a cooperation of insiders and knowledgeable outsiders.

## Private Organizations

Finally, the third type of organization in international scientific assistance and cooperation is that of the various private bodies, including private foundations, scientific societies (national or international), individual research institutions or educational establishments, and private industrial companies. These organizations usually have an easy access to the opinions of scientists about the cooperative programs, but beyond such relevant yet anecdotal evidence efforts are rare to assess programs. Routine annual reports of such projects tend to concentrate on input, listing the number of man-days of contact between visiting scientists and their counterparts or the number of people who attended meetings organized by the program. These are not irrelevant figures, but they are indicative only of activity and not necessarily of productivity or progress.

If a program is small, simple, and has one limited objective, evaluations are of course much simpler. The Physics Interviewing Project[39] is such a small program, generating information through personal interviews about students in developing countries who are applying to graduate schools in physics in the United States and other English-speaking countries in order to get an advanced education. The information is aimed at making easier the evaluation of such applications by respective graduate schools. This project, at least once since its beginning in 1969, attempted to evaluate itself by comparing the assessment of the interviewers with the actual performance of the student a year and a half later in the graduate school. Even in this simple case, problems arose owing to the inability to locate all students who were interviewed that year and to the incompleteness of the information that was obtained about their performances in the graduate schools. Furthermore, such an evaluation was done only once, while it should be done periodically. The fact that such assessment efforts must be done in addition to the regular daily activities of an organization is an additional deterrent to assessment.

## SUMMARY

The considerations in this chapter can be summarized as follows:

1. Assessments must be carried out in a value frame, and the first order of business is to specify the objectives of and the motivations for doing science.
2. The assessment of science is difficult, much more so than that of technology or production, because science has a broad range of objectives and motivations and because the product of science is something intangible—knowledge.
3. Existing assessments almost always deal only with input and rarely with output.
4. The assessment of scientific research in terms of its intrinsic criteria can be done quite successfully by a combination of peer review and scientometric methods.
5. The method of assessment described in number 4 is very rarely used in developing countries because of the unfamiliarity with the methods and the inability and reluctance to use peer review.
6. The assessment of the impact of scientific research on technology and on the aspirational, cultural, and cognitive dimensions of the country needs a well-defined and practical methodology, but even the existing ad hoc methods of evaluation are seldom used.
7. A methodology to assess international scientific assistance and cooperation has not been formulated, though a workable version could be created relatively easily.
8. There are strong disincentives in the structure of many organizations dealing with international scientific assistance and cooperation to engage in an assessment of their own activities.
9. Relatively speaking, the assessment described in number 8 can be done more easily in national and private organizations than in international ones and more easily in small and flexible organizations than in large and highly visible ones.

## RECOMMENDATIONS

Recommendations follow directly from the above conclusions.

1. Organizations and institutions in developing countries should be greatly encouraged to engage much more heavily in the assessment of science in their own countries and in the assessment of their international scientific collaborations.

2. To facilitate number 1 the existing methods of evaluation should be widely publicized in the developing countries, and assistance should be offered in the utilization of such methods.

3. The scientific communities in developing countries should be urged to be much more active and articulate in analyzing and then publicizing the problems of science in developing countries and the suggestions and proposals that could help in resolving those problems.

4. Developing countries should be urged to devote more attention to the quality of their scientific infrastructure and to deemphasize the overstress on purely quantitative considerations.

5. Developing countries should be urged to orient themselves in planning and other aspects toward indicators of output rather than of input.

6. Incentives are needed for much more research on methods of assessing science, particularly in the developing countries. With the support of the developed countries, it would be especially useful for scientists in the developing countries to develop, to do research on, and to generate indicators from programs that could be used in their own context.

7. There is a great need for the collection and utilization of data about the output of scientific efforts in developing countries that are not available through the large international data banks. Each country should be urged to engage in such collection of information.

8. Research is particularly needed to evolve methods of assessment for science as an aspiration and as a molder of our world view.

9. Methodology should be created, through a task force as outlined earlier, for the assessment of the work of organizations in international scientific assistance and cooperation.

10. The organization described under number 9 must be given incentives for engaging in self-assessment, either internally or, preferably, in cooperation with outsiders.

## NOTES

1. For a more detailed discussion of the concepts of science and its variants as well as of technology, see Moravcsik (1980a). Eloquent discussions, particularly of technology, can be found in the writing of Price, for example Price (1968).

2. Philosophers, as well as envious social scientists, frequently object to claims of objectivity by the natural scientists, since the former interpret such claims as having an "absolute," epistemological basis. Whether they do or not is not really relevant to the practical meaning of objectivity. It is enough, as Ziman (1968) so

eloquently explained, that the natural scientists have arrived among themselves at a consensus concerning what such an "objectivity" means in an operational sense.

3. For a discussion of the similarities and differences between the natural sciences and art, see Moravcsik (1974).

4. Paradigm is a conceptual framework for a scientific theory, a way of looking at a set of natural phenomena. For example, Newtonian mechanics, with the concepts of point masses, continuous trajectories in space, forces, and an absolute coordinate reference frame constituted such a paradigm. Long known by the scientists themselves but poorly articulated to outsiders, this concept was perhaps first explained in an eloquent and broadly understandable way by Thomas Kuhn (1970).

5. For a detailed exposition of why there is only one science, see Moravcsik (1978).

6. A very eloquent and instructive exposition of this historical dimension of science was given in Basalla (1967).

7. To take a recent example, the claim that there are several possible sciences was the basis of a recent meeting called Workshop on an Alternate Approach to the Management of Research, Development, and Education in Science and Technology, held in Bangalore, India, in June 1980. The two principal organizers of this meeting recently published a sizable treatise with this claim as the basis. See Shiva and Bandyopadhyay (1980).

8. For example, see another publication by the Bangalore group mentioned in note 7, namely, Seetharam, Shiva, and Bandyopadhyay (1980).

9. This trichotomy was discussed in Moravcsik (1976a) and in subsequent writings.

10. I readily admit that this brief discussion of the relationship between science and technology is semantically imprecise and sketchy. For a better try, see Moravcsik (in press).

11. A very eloquent, accessible, yet authoritative account of this is given in Clark (1969).

12. For a semipopular but very influential example, see Berne (1973).

13. The connection between space colonization and the evolution of science and technology in developing countries was discussed in Moravcsik (1980b).

14. I traced the similarity in attitudes between science and development in Moravcsik (1975).

15. This trichotomy of concepts was first discussed in detail in Moravcsik (1973).

16. This trichotomy was first outlined in a paper, Martin (1980), that is also significant for other reasons in the literature of assess-

ment, since it presents a beautiful demonstration (though not in developing countries) of what present methods of assessment can do in a comparative study of several scientific institutions in the same field. See also Irvine and Martin (1980).

17. This was also discussed in greater detail in Moravcsik (1975).

18. For example, see the volume on Greece, which was part of the set of national reports of the pilot teams, and also the volume in the same series on the evaluation conference in general (Organization for Economic Cooperation and Development [1968a, 1968b]).

19. For an excellent summary of the UNESCO approach to science policy, see United Nations Educational, Scientific and Cultural Organization (1979). See also Andrews (1979) for a UNESCO-organized project studying the effectiveness of research groups in the sciences in six small to medium-sized countries in Europe. The work was based primarily on a questionnaire distributed to some 11,000 members of some 1,200 research groups, though some scientometric indicators were also used. The main aim of the study was not to assess specific groups or individuals but to find factors that affect "effectiveness" in general. The results of this survey generally agree well with the conventional wisdom about such questions that reside in the scientific community itself.

20. For examples of OAS country assessment documents, see Organization of American States (1974a, 1974b, 1974c, 1975).

21. For four excellent country reports prepared by a first class team for AID, see, for example, Schweitzer (1979a, 1979b, 1979c, 1979d). The time allowed for the preparation of these reports was quite short and the resources limited. In spite of the high quality of the authors, the report does indicate the AID approach toward such assessments and the absence of output indicators. Schweitzer (1979c) is specifically oriented toward the formulation of a methodology.

22. For a review, see Moravcsik (1977a) as well as Meadows (1974) and Menard (1971).

23. Just to mention a few names (hence doing injustice to those left out), one might refer to the following: the copious writings of J. D. Frame and his collaborators F. Narin and M. Carpenter, for example, see Frame, Narin, and Carpenter (1977) and Carpenter and Narin (1980); the writings of H. Inhaber (1975); the articles by T. Eisemon (1979); a series of scientometrics articles by J. Vlachy (1979), some of which deal also with developing countires; the works by T. Braun and collaborators, some of which pertain to developing countries (Nagy and Braun 1980); and an article by Blickenstaff and Moravcsik (1982).

24. Again, just presenting a sample and hence ignoring some other notable contributors, let me mention the following: extensive studies by A. B. Zahlan (1980) on Arab science and technology, cul-

minating in an excellent book; the book on India by a group at CSIR Delhi (Rahman, Bhargava, Quarashi, and Sudarsan [1973]); a number of very informative articles by Arunachalam and Markanday (1981) in India; a quite unusual book on Pakistani science and technology (with general science policy implications) by Quarashi (1978); and an interesting analysis, with some scientometric sidelights, on Mexican science by Schoijet (1979).

25. The original exposition of this correlation is in Price (1969).

26. See Frame (1977) and subsequent writings by Frame.

27. This derivation is given in Moravcsik (1976b, pp. 152-54).

28. The original statement is in Lotka (1926). See also Price (1963, pp. 42-48).

29. For some discussion of applied scientific research in developing countries and of its interface with technology, see Moravcsik (1979, in press).

30. Scientific research as an overhead in the economic sense is also discussed often in the writings of D. deSolla Price.

31. Patents as a measure of output are used fairly frequently in the better science policy documents; see Rahman, Bhargava, Quarashi, and Sudarsan (1973, p. 132). This is acceptable if it intends to measure the output of the technological infrastructure. It is not necessarily appropriate, however, for the measurement of the effectiveness of applied scientific research.

32. Some of the interesting contributions in this area are by Anderson (1977), Lomnitz (1979), and Gasparini (1969). The last of these is a comprehensive survey of the Venezuelan scientific community through interviews and questionnaires.

33. I can personally attest to two occasions when representatives of such international organizations (in one case of UNESCO, in another of OAS) have stated publicly at international meetings on science policy that they considered the number of science councils and other similar bodies as a measure of the output of their organizations.

34. A very incomplete but perhaps representative list of such outstanding individuals who, for varying lengths of time, exercised a very favorable impact on their countries' scientific and technological development while working under the umbrella of such science policy bodies includes the following names: Bhabha in India, Usmani in Pakistan, Munroe in Buyana, Sabato in Argentina, Choi in South Korea, Baiquni in Indonesia, Wijesekera in Sri Lanka, Saavedra in Chile, and Roche and Villegas in Venezuela.

35. An example of an imaginative and helpful program is the informal communication and exchange network among Southeast Asian scientists that was originated by James McDivitt during the time he was the head of the Regional Office for Science and Technology for Southeast Asia of UNESCO.

36. For an extensive yet incomplete bibliography, see Moravcsik (1976b, 1977b). These two lists are, for example, definitely deficient on the Spanish language literature, which is produced especially in Latin America.

37. This occurred in a discussion between myself and F. Mayor, deputy secretary-general of UNESCO, in September 1978.

38. I want to give two examples. The first is the attempt by BOSTID of NAS to assess its own activities between 1969 and 1971, resulting in a substantial report that, however, was released only for internal distribution, hence, I am presumably not at liberty to list it as a formal reference. An assessment of the impact of BOSTID publications was included in a survey that I made of the Thai scientific and technological community, in cooperation with Yongyuth Yuthavong of Mahidol University, when I was a consultant in connection with the formulation of the Thai national paper for UNCSTD. BOSTID was also the manager of the well-known Brazilian chemistry program, which was also assessed, in a fashion, at its termination, but at a time when its real output and impact could not have been ascertained. The other example is the current survey of the activities of INT of NSF. A committee of outsiders has been formed to assess the record of INT, and a questionnaire is also being distributed to former "clients" of INT to get an opinion of the operation of INT programs. It is too early to assess the design, effectiveness, and impact of this assessment. IDRC also has some in-house assessment activities.

39. For information on the Physics Interviewing Project, see Moravcsik (1972) and Callen (1978).

## REFERENCES

Anderson, R. S. 1977. "The Government Scientific Institutions: Case Studies of Two Research Laboratories in the Late 1960's." Contributions to Indian Sociology 2: 137.

Andrews, F. N., ed. 1979. Scientific Productivity. Cambridge: At the University Press and Paris: United Nations Educational, Scientific and Cultural Organization.

Arunachalam, S., and S. Markanday. In press. "Science in the Middle-Level Countries—A Bibliometric Analysis of Scientific Journals of Australia, Canada, Israel, and India." Journal of Information Science.

Basalla, G. 1967. "The Spread of Western Science." Science 156: 611.

Berne, E. 1973. What Do You Say after You Say Hello? New York: Bantam Books.

Blickenstaff, J., and M. J. Moravcsik. 1982. "Scientific Output in Developing Countries." Scientometrics, vol. 4.

Callen, E., and M. D. Scadron. 1978. "The Physics Interviewing Project: A Tour of Interviews in Asia." Science 200: 1018.

Carpenter, M., and F. Narin. 1980. "The Subject Composition of the World's Scientific Journals." Scientometrics 2: 53.

Clark, K. 1969. Civilization. New York: Harper & Row.

Eisemon, T. O. 1979. "The Implantation of Science in Nigeria and Kenya." Minerva 17: 504.

Frame, J. D., F. Narin, and M. P. Carpenter. 1977. "The Distribution of World Science." Social Studies of Science 7: 501.

Gasparini, O. 1969. La Investigacion en Venezuela: Condiciones de su Desorrollo. Caracas: IVIC.

Inhaber, H. 1975. "Distribution of World Science." Geoforum 6: 231.

Irvine, J., and B. R. Martin. 1980. "A Methodology for Assessing the Scientific Performance of Large Research Centers." Proceedings of the Conference on the Evaluation in Science and Technology—Theory and Practice. Scientia Jugoslavica (Dubrovnik) 6, nos. 1-4: 83.

Kuhn, T. 1970. The Structure of Scientific Revolutions, 2d ed. Chicago: University of Chicago Press.

Lomnitz, L. 1979. "Hierarchy and Peripherality: The Organization of a Mexican Research Institute." Minerva 17: 627.

Lotka, A. J. 1926. "The Frequency Distribution of Scientific Productivity." Journal of the Washington Academy of Sciences 16: 317.

Martin, B. R., and J. Irvine. In press. "Assessing Basic Research: Some Partial Indicators of Scientific Progress." Research Policy.

Meadows, A. J. 1974. Communication in Science. London: Butterworths.

Menard, H. W. 1971. Science: Growth and Change. Cambridge, Mass.: Harvard University Press.

Moravcsik, M. J. In press. "Reflections on the Science-Technology Link."

_____. 1980a. How to Grow Science. New York: Universe Books.

_____. 1980b. "Preparing the New Pioneers." Bulletin of the Atomic Scientist 36: 56.

_____. 1979. "Linking Science with Technology in Developing Countries." APPROTECH 2: 1.

_____. 1978. "Do Less Developed Countries Have a Special Science of Their Own?" Interciencia 3: 8.

_____. 1977a. "A Progress Report on the Quantification of Science." Journal of Scientific and Industrial Research (India) 36: 195.

_____. 1977b. "Science and the Developing Countries." Position paper written for the U.S. National Science Foundation as an input into the U.S. National Paper for the United Nations Conference on Science and Technology for Development. (Copies obtainable from the Policy Research Analysis Division, Rm. 1229, NSF, Washington, D.C. 20550.)

_____. 1976a. "The Context of Creative Science." Interciencia 1: 71.

_____. 1976b. Science Development—The Building of Science in Less Developed Countries. Bloomington, Ind.: PASITAM.

_____. 1975. "Scientists and Development." Journal of the Science Society of Thailand 1: 89.

_____. 1974. "Scientists and Artists—Motivations, Aspirations, Approaches, and Accomplishments." Leonardo 7: 255.

_____. 1973. "Measures of Scientific Growth." Research Policy 2: 256.

_____. 1972. "The Physics Interviewing Project." International Educational and Cultural EXCHANGE, Summer, p. 16.

Nagy, J., and T. Braun. 1980. "Mennyire Nemzetköziek a Nemzeti Termeszettudományi Folyóiratok?" [How international are the national scientific journals?] Magyar Tudomany 25: 455.

Organization of American States. 1975. "Estudios sobre el desarrollo científico y tecnológico, No. 22: potencial científicotecnológico del Peru." Washington, D.C.: OAS.

_____. 1974a. "Estudios sobre el desarrollo científico y tecnológico, No. 17: aspectos cuantitativos de la ciencia argentina." Washington, D.C.: OAS.

_____. 1974b. "Estudios sobre el desarrollo científico y tecnológico, No. 18: inventario del potencial científicotecnológico de la Republica Dominicana." Washington, D.C.: OAS.

_____. 1974c. "Estudios sobre el desarrollo científico y tecnológico, No. 19: ciencia e industria—un caso argentino." Washington, D.C.: OAS.

Organization for Economic Cooperation and Development. 1968a. National Reports of the Pilot Teams, Greece. Paris: OECD.

_____. 1968b. Pilot Team Project, Evaluation Conference. Paris: OECD.

Price D. deSolla. 1969. "Measuring the Size of Science." Proceedings of the Israeli Academy of Sciences and Humanities 6: 98.

_____. 1968. The Difference between Science and Technology. Detroit: Thomas A. Edison Foundation.

_____. 1963. Little Science, Big Science. New York: Columbia University Press.

Quarashi, M. M. 1978. The Mechanics of R&D and Technology Transfer. Islamabad: Pakistan Academy of Sciences.

Rahman, A., R. N. Bhargava, M. M. Quarashi, and Pruthi Sudarsan. 1973. Science and Technology in India. New Delhi: Indian Council for Cultural Relations.

Schoijet, M. 1979. "The Condition of Mexican Science." Minerva 17: 381.

Schweitzer, G. E., and F. A. Long. 1979a. "Contributions to Industrial Development of Science and Technology Institutions in Malaysia, Nigeria, and Columbia, and Opportunities for Bilateral Cooperation (A Cross-country Comparative Analysis)." Final Report to the U.S. Department of State. Program on Science, Technology, and Society, Cornell University, Ithaca, N.Y.

_____. 1979b. "Contributions to Industrial Development of Science and Technology Institutions in Malaysia and Opportunities for Bilateral Cooperation." Final Report to the U.S. Department of State. Program on Science, Technology, and Society, Cornell University, Ithaca, N.Y.

_____. 1979c. "Developing a Methodology for Assessing the Contribution to Industrial Development of Science and Technology Institutions in Developing Countries." Final Report to the U.S. Department of State. Program on Science, Technology, and Society, Cornell University, Ithaca, N.Y.

_____. 1979d. "Contributions to Industrial Development of Science and Technology Institutions in Columbia and Opportunities for Bilateral Cooperation." Final Report to the U.S. Department of State. Program on Science, Technology, and Society, Cornell University, Ithaca, N.Y.

Seetharam, K., V. Shiva, and J. Bandyopadhyay. 1980. "An Alternative Perspective." International Foundation of Development Alternatives Dossier 20, nos. 11-12: 13.

Shiva, V., and J. Bandyopadhyay. 1980. "Science in India—Research without a Community." Bangalore, India: Indian Institute of Management.

United Nations Educational, Scientific and Cultural Organization. 1979. Science Policy Studies and Documents, No. 46: An Introduction to Policy Analysis in Science and Technology. Paris: UNESCO.

Vlachy, J. 1979. "Publication Output of World Physics." Czechoslovakia Journal of Physics B29: 475.

Zahlan, A. B. 1980. Science and Science Policy in the Arab World. New York: St. Martin's Press.

Ziman, J. 1968. Public Knowledge. Cambridge: At the University Press.

# 2

# FROM TECHNOLOGY ASSESSMENT TO AN INTEGRATED PERSPECTIVE ON TECHNOLOGY

FRANÇOIS HETMAN

Many obstacles continue to hamper the transformation of the technology assessment concept into a major social innovation. Persistent difficulties are seen on the procedural side. "Even after well over a decade of experience and the completion of hundreds of assessments, technology assessment (TA) is not an exact science."[1]

By definition, it never will be. Its primary concern is to look into the future consequences of today's choices, in other words to explore and, if possible, to reduce uncertainties inherent in the developments to come that are not amenable to factual knowledge but can be investigated with all possible scientific rigor.

Probably more important is the recognition that TA implies a fundamental change in attitude toward technology and a potentially far-reaching shift in the approach to decision making. Because of the experience gained, from now on it raises new questions of a less methodological kind and pertaining more to its function, status within the policy machinery, and use and usefulness in the formulation of policy options.

How effective is the change in the general attitude in government and in the public at large toward technology? Does it lead to a clearly acknowledged involvement at the institutional level? Is the commitment such as to stand up to the need for both clarifying the impacts of technological developments and gearing technology to socially beneficial applications?

The contention that "there is no theory of social change adequate to the requirements for rational-logical technology assessment"[2] does

The opinions expressed by the author are his own and do not necessarily commit the organization to which he belongs.

not mean that TA is subject to particular caution on scientific grounds, since it calls upon a specific and flexible mix of analysis, insight, and imagination. Social change—the object of all policy—is not amenable to nor based upon a social theory. It is a process originating in socioeconomic and sociopolitical moves, conflicts, and tensions provoked by contradictory decisions of people with different interests and ambitions.

This is precisely what raises the need for technology assessment as a systematic aid to better informed policy making, attempting to discern ex ante possible patterns of developments instead of just categorizing ex post experience.

## FRUSTRATION AND HOMEOSTASIS

Scientific research and technological innovation are basic determinants of social change. In our technological society, they are factors of enhanced performance but, by the same token, causes of destabilization. They create new areas of uncertainty. They make whole categories of knowledge, technique, and professional skill obsolete. They call into being new risks and challenges.

These are man-made. No external or supranatural correcting mechanisms are available. People have to come to new terms with technology-induced human systems. This implies finding new ways of bringing technological innovation and the proliferation of technologies under some kind of social control.

This runs counter to a broadly accepted view that technology is a self-enhancing factor that sweeps over society in an irresistible way. According to this "technological determinism," technology is advancing autonomously, whereas human adaptability and social structures are stagnating and being left behind. This discrepancy provokes a feeling of frustration and growing alienation, which in turn, accentuates the lack of confidence in existing social institutions.

At the same time there seems to be no other way out than to look to science and technology for remedial action. The expansion of research and development (R&D) activities on a large scale and the establishment of institutions for science policy raise hopes that problems can be overcome within a more consistent science-technology model producing its own homeostatic rules.

However, science-fueled and technology-fueled technological change continuously reveal incompatibility or conflict between old objectives, routine, decisional patterns, and new issues. This raises the need for new relationships among research, technological innovation, and decision making, taking into account not only technical and economic performance parameters but also the societal dimension of impacts of new technological developments.

Two main trends can be observed, divergent if not contradictory, as far as objectives and policy implications are concerned. On the one hand, there is a tendency to strengthen the role played by R&D activities in tackling problems raised by economic, social, and political uncertainties that prevail all over the world. In most countries, science and technology are expected to contribute more significantly to enhancing the competitiveness of the national economy and to limiting dependence on external sources of energy, raw materials, and products. National authorities take measures to optimize the allocation of R&D resources through sounder program management, stricter control of means, and a more rational use of research findings.

On the other hand, there is a call for a greater social relevance of science and technology. So far this has led to only a shifting of research priorities. Generating research of societal relevance raises the extremely complex and elusive question of identification, selection, and pursuit of social goals. This touches on issues of values and social consensus, on the one side, and on political concern, will, and long-term commitment to gearing science and technology systems to new directions, on the other.

A diffuse phenomenon is the public concern about science and technology. Various social groups claim social and political competence to have their say on policies that have traditionally been regarded as the exclusive province of scientific and technological expertise or dynamic entrepreneurship. Such challenging is expressed in the technology assessment movement among others.

## CONSEQUENCES FIRST

Thus, the concept of technology assessment leads to a reappraisal of the role of science and technology in contemporary society. It implies a much deeper understanding of the nature of technology and technological innovation, a careful consideration of the consequences of alternative technological decisions, and a more open and broader approach to decision making.

Technology assessment can be defined as a process of analysis, forecasting, and assessment of technological futures and their impacts on society resulting in action options for decision makers. On the analyst's side, it encompasses the study of technological parameters; the elaboration of technological forecasts; the analysis of social, environmental, cultural, and political factors; a general assessment of all relevant effects and possible consequences of a technology; and an evaluation of alternatives. On the decision maker's side, it implies appropriate institutional mechanisms that make it possible to identify demands for technological change, gear scientific and technological knowledge to societal needs, make the choice of socially acceptable technologies, determine suitable means of action, and plan the appropriate phases of implementation.

Since the beginning there have been two tendencies diverging in approach and philosophy. The first tendency can be called the technologist's approach. It considers that technology assessment examines the impacts of all alternative policies that can be followed but does not come out with a specific recommendation involving value systems. Assessment ends at the frontiers of the broadened technological analysis, leaving the rest to traditional, existing social and political processes. The second tendency, on the contrary, regards the management of technology as a part of overall planning or "social engineering." In its extreme formulation, it starts by spelling out values, social policies, and objectives and works down to technology assessment in order to clarify the most appropriate technical options.

These two points of view may be considered as extreme lines of definitional mapping between which technology assessment can take a great variety of forms. They are not mutually exclusive but rather complementary as they address themselves to the same cluster of problems but from a different angle and at a different level of the decision-making process.

In the narrow sense of the term, environmental impact statements can be qualified as limited TA's centered on one category of impacts, namely, the environment. They represent a form of assessment institutionalized in the United States with the enactment of the National Environment Policy Act. This act requires that all federal agencies include in every proposal for legislation or actions affecting the quality of human environment a detailed statement on the impacts of the proposed action. Similar legislation and procedures have been adopted in a number of other countries together with the establishment of government mechanisms for environmental protection.

On the contrary, integrated assessment, such as was adopted by the U.S. Department of Energy (Energy Research and Development) for its programs, comes as close as possible to the ideal of comprehensive assessment. This is expected to shed light on impacts of new technologies developed by the R&D branch of the department both from the point of view of society and the environment. It does not include the term technology, which seems to imply that the emphasis is on the analysis of impacts and consequences of a given political choice and that the concept is applicable to decision making in general.

The concept of systems assessment of new technology (SANT) has been outlined within the International Institute for Applied Systems Analysis (IIASA) as a part of its investigations in the fields of systems analysis and technology management. It is represented as a framework for the comprehensive study of the whole set of analytical efforts, ranging from technological forecasting to technology assessment, alternative technologies, evaluation of R&D, and science-technology potential. The object of SANT is organized technology, a more normative approach to both technological and organizational innovation.[3]

To convey the idea that technology assessment involves basically a change in attitude toward technology, a range of other terms was suggested, such as study of impacts, analysis of social impacts, social management of technology, program impact evaluation, participatory technology assessment, and so on. The same concern has led the Organization for Economic Cooperation and Development (OECD) to adopt the expression "social assessment of technology" to distinctly point out that the main thrust is on the systematic analysis of impacts on society of new technological developments and their diffusion.[4]

Reviewing the experience of the U.S. Office of Technology Assessment (OTA), its director, John H. Gibbons, says that "technology assessment is best understood and employed as embracing a variety of analyses on how to resolve sociotechnical problems. Technology assessment is a process through which scientific methods are used to understand how society can develop, influence, and apply technology to meet social needs in the most beneficial way."[5]

The U.S. National Science Foundation (NSF), which is probably the body with the longest and broadest experience in TA studies, distinguishes four stages in its efforts in this field, including cost-benefit analysis, comprehensive identification of effects, policy-oriented analysis, and a combined program of technology assessment and risk analysis. Presently, NSF considers that the primary objectives of technology assessment are to "1) identify and analyze the relevant economic, social, technological, legal, psychological, institutional, environmental and political consequences of a projected technological change; 2) analyze the ability of existing institutions to accommodate to institutional change; 3) explicate the different points of view of the parties at interest; 4) compare the alternative policy and technological choices available to decision makers; 5) identify and analyze the uncertainties and risks associated with alternative policy choices."[6] These are very close to the five main characteristics of social assessment of technology as described by the OECD.[7] Thus, experience confirms that to be a valid policy tool, technology assessment must be organized as a continuous activity, as a process building both on facts and foresight, and as a permanent interplay between decision makers and analysts.

## APPLICATIONS

Ideally, technology assessment should cover the whole range of issues raised by the multifold interfaces between technology and society. It should examine all virtualities of a technology and sift out the most beneficial alternatives. According to the nature of each interface and the developmental stage of a given technology, TA can be called upon to illuminate corrective action or to explore the needs and conditions for new, "socially desirable" technological developments.

### Monitoring of Negative Effects of Technology

Undesired effects of a technology are frequently of a nonlinear type. They appear in a cumulative way and create a new collective need to abate and master them. There is a "threshold effect" both in physical terms and in the degree of public awareness. Therefore, TA should not only map out the future development of a technology but also identify the thresholds under which should be kept undesirable effects. Though this sounds relatively easy, it may be extremely difficult to detect phases of expansion and breaking points that lead to new unexpected branchings.

### Abatement of Environmental Disruption

In a sense, social assessment of technology relative to the environment is the reverse of the evaluation of a nascent technology. Instead of starting from a technology and looking after its effects and impacts, one begins with a stocktaking of impacts and proceeds to the causes of a technological nature to show processes that end up in having unfavorable consequences.

The government decision makers look for criteria that will enable them to optimize the well-being of society after having taken into account the effects of the various socioeconomic activities on the environment. They tend to break down the quality of the environment (considered as a whole) into a series of specific questions. Courses of action are established for each case, and advantages or disadvantages of each course of action are examined with regard to an optimal solution.

### Screening of Potential Developments

Such a screening includes the identification of research areas most likely to generate technological breakthroughs, the detection of those technological developments that are likely to lead to important innovations, the identification of possible innovations that seem more efficient and more desirable on social grounds than existing technologies, and the definition of criteria for evaluation of the desirability of such possible innovations.

The available experience and empirical studies show that the relationship between basic knowledge and technological innovation is much more complex than has been admitted by the prevailing sequential thinking, according to which technology evolves gradually from scientific knowledge. Screening of potential new technologies requires, therefore, a particularly imaginative approach, since it is supposed to take into account a large number of uncertainties. It can be envisaged as a continuous search coupled with sensitivity analysis to shed light on different parameters of each new technological option.

## Research of "Socially Desirable" Technologies

If technology can be considered as a dependent variable, not self-justified but serving fundamental social objectives, the starting point of TA shifts to social goals. The crucial problem is to define the objectives of society so as to derive from them groupings of socioeconomic programs and subsequent technological needs.

This goal-oriented approach necessitates a policy conversion of social goals into operational objectives, and, still more important, a workable consensus on such objectives. These conditions are hardly met in the real world. Society is still ill-equipped to handle conflicting interests, and social groups may be unable to agree upon the criteria for a choice of technologies that would lead to the achievement of objectives that are continuously blurred by the complexity of socioeconomic phenomena.

It is understandable, therefore, that most TA studies are of a "corrective" type, mainly as a direct response to questions concerning particular categories of effects or risks of already known technological developments on the eve or at some early stage of application.

# CONSTRAINTS OF ANALYSIS

Critics of TA contend that only a very few studies have approximated the initial ambition or expectation. There are numerous reasons for that. One of them is the unrealistic delineation of the study in comparison with constraints of information sources, timing, money, and available skills. Another one is the discrepancy between the awareness of the sponsoring agency and its ability to make use of findings in formulation of policies.

The need to assess the technology assessment movement and available experience stimulated a number of analyses. Leaning on the most comprehensive ones,[8] some general remarks can be made on points of fundamental importance: evaluation of impacts and formulation of action options.

## Evaluation of Impacts

The definition, identification, and evaluation of impacts comprise a central part of TA. It should, therefore, be the most accomplished part of the study. In reality, it ends up being the most elusive, in spite of many-faceted and often-enlightened efforts to supply relevant information on discernible effects of a technology.

In most studies, impacts other than technical and economic are framed separately in broad and general terms. Frequently, impact areas are identified on a conceptually preselected basis, inferring them by analogy from the past. The result is that some categories of effects receive particularly insistent attention, whereas whole categories of other impacts are ignored. Since no reasons are or can be given for selecting one category of impacts rather than another, the general impression is that the coverage of impact areas is highly intuitive and often dictated by a personal scale of preferences or by the degree of accessibility to readily usable data.

The difficulty of identifying social areas likely to be impacted is compounded when it comes to an evaluation of the impacts. This is the hard core and the most perplexing part of the analysis. In many studies, logical or imaginative inference is used to draw up a comprehensive array of possible impacts. However, such an extensive listing implies running a risk of being merely a formal brainstorming exercise if it is not accompanied by an effort to shed some light on the magnitude and intensity of impacts.

In general, evaluation concentrates mainly on technical and economic factors and aspects. Other categories of impacts are treated in descriptive and more or less peripheral terms. There is a clear-cut methodological break between the former group of effects—demand amenable to statistical and numerical evaluation—and the latter—considered as being of a distinctly qualitative nature.

The final outcome is heavily unbalanced. Given the inherent or supposed difficulties of quantification, the attention is focused on quantifiable elements, while the puzzling problem of attaching some commensurable value to social aspects is pushed yet farther away into the realm of analytical intractables.

This makes it hardly possible to think of drawing together the various impact categories into an integrated overall balance sheet of societal advantages and costs. The discussion goes on whether attempts at deriving some kind of social net value are sensible or advisable. The absence of such a summing-up rating is a major drawback for users of TA who may be easily inclined to doubt its realistic character and its usefulness to the decision maker.

## Formulation of Action Options

Considerable attention is devoted to the formulation of action options or suggestions for policy measures. Such extensive treatment may be surprising when one considers the rather inconclusive analysis of societal categories of impacts. This may reflect a more or less conscious effort to compensate for the somewhat frustrating part of the analysis dealing with evaluation of impacts. In many cases, an impressive list of suggestions is drawn up like an inventory of things that should be done rather than a truly rigorous definition of action options derived as closely as possible from the results of the assessment of impacts.

There is a general tendency to give as many suggestions or recommendations as possible. These suggestions frequently resemble an enumeration of desiderata formulated without any attempt to establish a hierarchy or to order them by some rating based on a set of criteria of internal coherence and political feasibility.

The types of action most often investigated are legal norms and regulations. This is presumably because these are reputed to have the most general validity and because it is believed that they can be most easily acted upon by the public decision maker. Though some thought is given to issues that may arise as a consequence of actions or behavioral attitudes of the interested social groups on the one hand and the affected parties on the other, the general tendency is to avoid these questions in final suggestions or policy recommendations.

This shows how difficult it is to formulate a range of action options set in a consistent political framework so that they could unequivocally impose themselves on the attention of decision makers. There is little evidence that recommendations resulting from TA studies have facilitated or simplified the decision-making process. Their main contribution seems to have been a systematic structuring of facts, a disclosing of uncertainties, and making policy makers aware of often unnoticed or neglected impact areas, as well as of concerns of the parties affected.

## INSTITUTIONAL PROBLEMS

As a new concept and a new approach to technological change and society in general TA has still to find its place within the institutional setup and decision-making mechanisms.

Where does it belong? Is it an outgrowth of science policy? Is it a new branch of general policy stemming from a reaction to the disenchantment with science and the subsequent questioning of the ends of scientific activity and the passive acceptance of technological developments?

It seems obvious that it should be linked to technology policy. There has not been such a thing so far. From the institutional point of view, technology has no place of its own. The word technology was only recently added to that of science in the names of both national agencies and international organizations. This marks an increased concern for matters of technology, but the "couple" of science and technology remains ill matched. A much deeper understanding of the interplay between technology and society is necessary before technology policy can be outlined in relevant operational terms without stifling free entrepreneurship.

With the exception of the United States and its Office of Technology Assessment (OTA), there is no special institution for TA. The major obstacle to TA is precisely the absence of a technology policy and a government body responsible for it. This can be illustrated by the experience of the OECD and its program on the social assessment of technology. The second phase of this program was intended to test the concept of TA in applying it to real problems of interested member countries. From a great number of topics suggested by national authorities, three subjects were selected for studies in common. Finally, a sufficient commitment was secured for the theme: new urban transportation systems. The project was carried out within a time span of about 18 months, following the exploratory stage devoted to defining the practicability and operational steps of the methodological approach and to the identification of national sources.

As a result, six national contributions were received and served as a basis for the final synthesis.[9] Most often they were parts of broader projects undertaken within the current program of the respective national agency. In some cases, they were merely derived from an already completed work carried out with a quite different objective in mind. This made it particularly difficult to identify common areas of interest on which a meaningful comparison could be made. In general, an effort was made to introduce the idea of the social assessment of technology either as an additional category of conceptualization or as a new tool allowing for the exploration of aspects until now neglected in traditional technical feasibility studies.

This shows that in the present situation it is extremely difficult to find for any candidate subject for TA the appropriate institution willing, able, and authorized to undertake such work. There is a wide dispersion of competencies among the various ministries and agencies. Even if interested in TA, they are hampered by their narrowly defined responsibilities. They are hardly inclined and often reluctant to modify their programs in order to devote some resources to TA.

As a rule, technology is developed as a trial-and-error process with opportunities for feedback from social, economic, and legal institutions. A feedback through TA should allow technologists to reshape a technology and to better adapt it to social objectives. However, the institutional inertia hinders both propagation and implementation of TA. This is not only for reasons of structural rigidity. Social consequences of a new technology are difficult to clarify and impossible to attest to objectively. Furthermore, business and political involvements make it generally hard to modify the course of action. This makes decision makers hesitant to rely on TA and to challenge ongoing technological trends.

## RESPONSIVENESS AND USEFULNESS

There seems to be a deliberate tendency to mix up science and technology within a common framework, a single system functionally attached to a distinct area of government policy. However, the boundaries of such a system are difficult to delineate, not only because of its intrinsic complexity but also because science and technology pervade all aspects of society. This places science and technology (S&T) policy in a pivotal but uneasy position: whereas it is looked upon as an essential political function, it depends widely on ends and directions pursued in other sectors. Accommodation of science and technology within the same policy machinery is based on a very simplified sequential model: science and technology are complementary parts of a single spectrum. Science produces new knowledge and technology applies knowledge to create new products and new sets of instrumental know-how.

There is, in the long run, a logical convergence in the utilization of knowledge in a historical perspective. But this is a result of processes that originate in different institutions and—still more important—that are inspired by different social groups and controlled by different economic interests.

Science and technology policy is a hybrid form of R&D policy (that is, organizations and measures in support of intellectual activities connected with scientific discovery, on the one hand, and with the inventive development of new ways of doing things, on the other). Research and development effort has been associated with a fast-growing number of societal issues. The magnitude of resources demanded has grown in accordance, so that the allocation of resources compels selection, priorities, and political choices. The euphoria of unrestricted expansion has been replaced by a target-oriented and programtied attitude. Instead of awaiting automatic returns to R&D expenditure, the emphasis is on making its results usable and on ensuring that they are really put to suitable use.

In other words, the thrust is on making R&D more responsive to other areas of government policy and, in particular, to demands for technology developments likely to enhance competitiveness and economic performance. However, here government policy makers move into an area where they have only limited control of key variables and rarely sufficient knowledge and understanding of market forces and potential consequences of new technological developments. Most frequently, they just react to some deleterious aspects of a technology in a later stage of diffusion. But these are cases of direct government intervention with the aim of creating new technologies or implanting whole branches of industries based on advanced technological systems, generally within broader policy schemes—baptized restructuring, redeployment, reindustrialization, innovation policy, and so on.

Such a piecemeal approach shows that government policy makers are ill-equipped to deal with technological change. Recently created institutions may have helped in defining priorities, but they do not by themselves guarantee either success or the desired follow-up. The experience of the OECD countries leads to a serious caveat: "While science and technology must indeed be more strictly oriented than in the past, we must be careful not to sterilise them by stifling them in a bureaucratic strait jacket. If it is to be really 'useful', research must be fully associated not only with the execution but also with the formulation of the policies it has to serve."[10]

Instead of clarifying the respective role of each of them, the formal association of science and technology accentuates the dichotomy and confuses the issues. The long-term function of science, inherent in the characteristics of scientific investigation, is very often at cross-purposes with the urgency of the short-term economic and social problems arising in all areas of government policy. The effect of such urgency is that science policy is being pressed for prompt answers, although the expected answers are essentially of an instrumental and technological nature.

This discrepancy has not been bridged by the new institutional mechanisms and forms of action introduced by most of the industrialized countries during the 1960s and 1970s. The experience of the OECD countries seems to show to the contrary the need for a clear distinction between research policy and technology policy.

A number of interesting attempts have been made to find appropriate modes for integration of R&D policy in general policies. These attempts include departmentalization, problem-oriented programs, and a sectoral approach to integration within general planning. On the one hand, there is the tendency to decentralize responsibilities at the operational level; on the other hand, emphasis is put on a hierarchical articulation of R&D activities at the level of centrally defined government objectives.

Whatever the institutional arrangements, the basic motivations boil down to directing the R&D effort toward stimulating the national capacity to develop new processes and products in order to create technology-intensive industries within a technology-impulsed economy. The general endeavor is to make R&D activities in economic and budgetary terms. The real justification is technology. Why is it not made explicit?

One reason is the difficulty of reversing the sequence and making science dependent on technology, inducing scientists not to pursue other research paths but those derived from technological objectives. Another reason is the absence within government and policy machinery of relevant criteria to direct new technological developments to clearly defined societal goals. The missing link is technology policy. However, technology policy is impossible without a firm basis for understanding functional relationships between technology and society.

Here steps in technology assessment, a discloser of issues and a consistent foundation upon which it should be possible to build up for action. Only a continuous process of screening technological possibilities, anticipating their development and pace of generalized acceptance, exploring their impacts on society, evaluating their consequences, and suggesting policy options can create basic conditions for socially significant technology policy.

## INTEGRATED PERSPECTIVE ON TECHNOLOGY

The most striking phenomenon of our times is the universal belief in the power of technology. Its corollary is a generalized public awareness of its promises and risks. Both in industrialized and developing countries, it has become a central concern of populations. In the first ones, technology is considered as the main if not the only factor of competitive advantage where continuous innovation can and must take place to adapt economy and society to planetary competition. In the latter ones, it is looked upon as a pivotal factor of development making possible a decisive move toward catching up with more advanced countries.

Surprisingly, the so-called disenchantment with science makes science, not technology, responsible for what is felt as a betrayal and deviation of technological progress. Charging remote manipulators of knowledge, it has not shaken the faith in the remedial virtue of technology. Whereas some specific applications of already existing technologies are feared and emotionally opposed, new technologies are called for even more strongly to bring more suitable technical solutions.

Can TA be instrumental in finding the right way to reduce negative effects and to optimize beneficial aspects of technology? Only if it leads to technology policy based on some operational criteria of what should be the net benefit to society of new technological developments.

Up to now, TA has evolved as a report-producing exercise without a direct connection to decision making. This is the case even for the OTA within the U.S. Congress. Conceived of as an early warning system, it does its best to answer difficult questions. While it finds some listeners, there are few institutional receivers to implement its findings. This makes it into a distinguished body, standing alone and aloof.

The unsuccessful attempts to define an appropriate institutional format for technology assessment activities and the rejection in West Germany and France of proposals to create within the orb of parliament a body similar to the OTA add to the skepticism with regard to introducing efficient assessment mechanisms without a decided policy commitment.

As far as Western European countries are concerned, it is often contended that ministerial departments involved in matters of technology have for a long time been used to pay attention to social implications of new technological developments, particularly in liaison with the increasing governmental involvement in promoting advanced technologies and in stimulating technological innovation.

In the new era of planetary competition for markets on the one hand and for resources on the other, governments are in need of a much integrated perspective on technology that would include an in-depth analysis of new technological developments and their societal implications so as to provide a rational basis for choices in the field of technology and to lead to an efficient monitoring function. This can be illustrated with a few recent examples.

Reviewing the unique experience of the NSF, the experts charged to pursue work in the field of TA consider that TA enters a new phase where modalities to meet the need for a historical perspective of TA should be developed. "In this way, the assessment process might eventually come to be institutionalized as an interactive problem-solving and learning process. The expectation is that successive assessments building on prior results would arrive at increasingly better approximations to the definitive insight desired."[11]

For the authors of the exhaustive inquiry into technological change in Australia there is a diversity of perceptions of technology assessment, but "the central issue appears not to be whether technology should be assessed and controlled, but the extent to which it should be assessed and controlled, and by whom. That is, what is at issue is the quality of decisions about technology and whose interests are taken into account when those decisions are made."[12]

More explicitly, the West German authorities put assessment and acceptance of technology among the priorities of their R&D policy.

> We must ensure that the future debate on the introduction and utilization of new technologies is not marked by a polarization of opinions incompatible with objective discussion and by a resulting loss of differentiation in the arguments and solution approaches presented. For this reason, it will be essential to conduct an objective discussion on the advantages and disadvantages of new technologies, before their large-scale technical or commercial use becomes imminent and before "practical constraints" create facts ahead of political decisions. In addition, government authorities will have to pay greater attention to the consequences of technologies and to increasing their powers of evaluation, and the development of technological alternatives must be initiated at an early stage, even if their utilization appears less promising for the moment.[13]

All these new approaches raise the question of the social or societal management of technology. What can be the types and modes of such a policy within industrialized countries, where the innovative process is mainly in the hands of individual organizations, which tend to make it an integral part of their global strategy of competition? Where can it be located to be influential without stifling innovation, to be effective without substituting government intervention for a dynamic pluralistic pattern of initiatives?

## DEVELOPING COUNTRIES AND TECHNOLOGY ASSESSMENT

It is frequently argued that developing countries are in some respect more free than advanced countries to foster their economic growth, without the sequel of disadvantages that are now becoming visible in the highly industrialized societies. How far does this assumption apply?

In the already overabundant literature on development, there seems to be a broad unanimity on the prominent role of technology. The order of the three terms of the simplest production function capital-labor-residual factor (including technological change) is practically reversed and reads as technology-capital-labor. Technology is considered as a driving force and a determinant of not only the rate of growth but the very pattern of development. It is viewed, from the standpoint of developing countries, as a pool of technical choices within which it seems possible to choose (for the same economic cost) a technological option that permits one to maximize the growth rate of development.

For the most part, this pool lies outside the immediate reach of developing countries. The rate and the pattern of technological advance are determined by industrialized countries and, in particular, by a few that can devote considerable resources to R&D and the organized exploitation of scientific knowledge in experimental development, new production processes, and the creation of new products. Consequently, technology must be imported from industrialized countries and adapted to local circumstances.

This transfer of technologies is a very complex phenomenon. Because of the deep and multifaceted differences between industrialized and developing countries, it tends to be viewed by the latter ones as a unilateral flow bringing a cumulative advantage to industrialized countries.

Policy makers in developing countries resent it as a limitation of their decision-making capacity. Thus, dependence on foreign technology is readily equated with an overall state of dependence. This reinforces the feeling of distrust as to the intrinsic constraints of an economic, social, and political nature that are brought in by the transfer of technologies. According to the well-known argument, imports of technology absorb scarce foreign exchange, foster capital-intensive sectors, and introduce noncongenial patterns of consumption. This results in a demand for more foreign technology and perpetuates the vicious circle of underdevelopment.

Thus, there are ideas of appropriate, intermediate, labor-intensive, or specific technologies for developing countries. In reality, the choice is constrained by the competitive advantage: this is in favor of technologies that entail the most important gain in productivity, which is the case of research and capital-intensive techniques launched in industrialized countries.

Specific techniques of basic importance can be invented only for a specific type of civilization. This would imply that there is another model of development that can be followed independently so to speak by developing countries. The generous theory of intermediate technologies seems to overlook the simple fact that the concept of underdevelopment is meaningful only in relation to developed countries; in other words, a technological society is the same model of reference both for developed and developing countries.

If developing countries cannot have a specific model of technological development, can they have their own technology policy? They dispose of only limited scientific resources and research facilities. Since technological innovation is concentrated in industrialized countries, technology policy of developing countries can bear mainly on the choice of imported techniques and their optimal utilization.

To reduce their technological dependence, developing countries are likely to grope for a combination of several possible policies: selective legislation, diversification of supply sources, introduction of new skills in technological packages, recruitment of external experts to create domestic facilities, and so on. Technology assessment can play a decisive role in formulation of such a mix of technological policies, provided that these policies are closely and organically linked to national objectives and available sociopolitical means. "The developing nations should give precedence to assessments of those technologies that strongly relate to basic needs and national goals. In the developing countries, TA at the initial stage would help to identify early and accelerate the exploitation of appropriate technologies that might otherwise not receive due attention."[14]

However, there are serious caveats and real difficulties. Technology assessment implies a systematic search for impacts and, in particular, for nonobvious long-term consequences. This requires a broad-based, future-investigating activity to map out possible futures and areas of uncertainty. Hence, there is a need for data and information on historical trends, linkages between socioeconomic and technological phenomena, possible modifications of production and consumption patterns, and so on.

To identify and compare the beneficial and harmful effects of a technology, it is necessary to project the path of its development and application within a specific societal system to determine both its contributions and critical thresholds. This requires a good stock of information on the diffusion of new technologies and the rate of public acceptance.

To take into account the social impacts of a technology, it is necessary to investigate the workings of social mechanisms, to determine conflicting interests, and to make an inventory of social groups likely to be affected by a given technological development. With respect to these and other aspects of TA, the scarcity of reliable data in developing countries makes it extremely difficult to evaluate technical and economic potentialities. Moreover, to apply a goal-oriented approach to technology policy and, by the same token, in TA implies a frame of reference—a model of societal development—within which the choice of technologies can be linked to the desired type of societal achievement. However, a clear-cut distinction between ends and means is hardly possible. The complexity of societal phenomena is such as to preclude the formulation of a detailed long-term, policy-relevant model of society. In the best case, only temporary objectives can be defined.

Technology policy requires creative TA to clarify available technological options, and it needs creative policy makers to get it in line with socially relevant issues.

NOTES

1. U.S., Round Table Discussion Paper, "Assessment of Technological Systems," CCMS Report no. 115, North Atlantic Treaty Organization, Committee on the Challenges of Modern Society, p. A-8.

2. Donald N. Michael, "Technology Assessment in an Emerging World," Technological Forecasting and Social Change, no. 11 (1978), p. 190.

3. G. M. Dobrov, "A Strategy for Organized Technology" (IIASA Workshop on Systems Assessment of New Technology: International Perspectives, Laxenburg, Austria, 1978).

4. Organization for Economic Cooperation and Development, Methodological Guidelines for Social Assessment of Technology (Paris: OECD, 1975).

5. U.S., Congress, Hearing before the Subcommittee on Science, Research and Technology, 96th Cong., 1st sess., no. 87, 10/10/1979, pp. 23-24.

6. Joshua Menkes, Technology Assessment: A Brief Overview (National Science Foundation, January 1, 1980).

7. Organization for Economic Cooperation and Development, Social Assessment of Technology (Paris: OECD, 1978).

8. U.S., Office of Technology Assessment, Technology Assessment in Business and Government, January 1977; Stanford University, Department of Engineering, "Economic Systems: Strategies for Conducting Technology Assessment" (Conducted by Joe E. Armstrong and Willis W. Harman, Stanford, December 1977); Mark R. Berg et al., Factors Affecting Utilization of Technology Assessment in Policy-Making (Ann Arbor: Center for Research on Utilization of Scientific Knowledge, Institute for Social Research, University of Michigan, 1978); Organization for Economic Cooperation and Devel-

opment, Social Assessment of Technology: Review of Selected Studies (Paris: OECD, 1978); and Vary T. Coates, Technology Assessment in Federal Agencies, 1971-1976 (Washington, D.C.: George Washington University, 1979).

9. Organization for Economic Cooperation and Development, Social Assessment of Technology in the Field of New Urban Transportation Systems (Paris: OECD, 1978).

10. Organization for Economic Cooperation and Development, The Research System, vol. 3 (Paris: OECD, 1974), p. 181.

11. Jiri Nehnevajsa and Joshua Menkes, "Technology Assessment and Risk Analysis," National Science Foundation, October 1980, p. 13.

12. Australia, Report of the Committee of Inquiry into Technological Change in Australia 1980, Technological Change in Australia, vol. 4 (Canberra: Australian Government Publishing Service, 1980), p. 376.

13. Der Bundesminister für Forschung and Technologie, Bundesbericht Forschung, vol. 6 (Bonn, 1979), p. 18.

14. United Nations, Technology Assessment for Development (New York, 1979), p. 6.

# 3

# TECHNOLOGY AS COSMOGENESIS: PROCESS TE(CHN)OLOGY

PAOLO SOLERI

## IS TECHNOLOGY A BIAS?

What technology is may not be an idle question. The following two considerations are being entered here to help answer the question.

1. The biological organism exists and performs via a technology of the flesh, a "biotechnology" that is not only immensely subtle but also premonitory of the much younger technology epitomized in homo faber's manipulation. A good illustration is the minimotor (electrical?) propelling the cilias of some microorganisms.
2. We are finding out that millions of years before the advent of the human animal exobiological techniques were developed by insect and other organisms—for example, farming by termites and ants, use of sticks and other objects by birds and mammals, and so on.

In fact, from an epistemic point of view, it can be held that a biotechnology of immense subtlety is present in all organisms and the technology of which we humans are the quasi-fiendish exponents and promoters is of the most recent origin. Pursuing this line of reasoning, we could begin to put some order in our historical archives and develop a more congruous chronology of things, somewhat as follows:

1. The megatechnology of the cosmos—the one classical physics has been trying to decipher;
2. The microtechnology of subparticles (wave reality)—the one in which new physics is entangled, in its efforts to understand its own earlier misunderstanding;

3. The microtechnology of "tangible" reality made of things subject to thermodynamics that control and unwind (entropy) the planet and its realities of the living progression;

4. Technology of the living stuff—the biotechnology of flesh, blue, green, and red peaking for now into the reflective technology, the technology of thought;

5. The exotechnology that flesh-mind has been inventing to foster its secret follies; this is "technology proper"—the one to be assessed in this book;

6. Religions—those mental/technological constructs attempting to give a meaning to the whole (literally) blasted thing (big bang) and doing so by a forward look concealed in endless backward-looking glances; the metaphors and the myths of religion developed by "te(chn)ologians";

7. Aesthetics—the twin sister of religion, the humble undertaking of which, the work of art, produces nodules of transcendence, the study divine, the creations of the technologist of beauty;

8. Last, the young latecomers of science and philosophy—the infants explaining the parental nexus and the cosmic clan, attempting by the torturous paths of shamanism, astrology, alchemy, exorcism, and the postulate of objectivity the interpretation and the understanding of the whole (blasted) thing.

In this chronology, technology is the deus ex machina. All that comes about and is in the process of becoming technology. In this definition, technology is time itself.

As one can speak of prototechnology (numbers 1, 2, and 3) as the cosmic stuff going about in quasi-probabilistic ways, can one speak of posttechnology? Perhaps yes, if one could foresee the convergence of biotechnology (the brain as its fruit) and of the supertechnology of mechanical, chemical, and microcircuitry under the spell, the exorcism of mind-spirit. The ensuing product could be the divine technology of grace-equity and beauty. This hits at the question that has begun to obsess the best minds of our time, namely, what is the future of flesh, love, grace, sex, beauty, and so on? Can these priorities be transferred to a silicone mind? Of course, mind is not flesh nor silicone, but it is how flesh speaks to itself and to its "otherness" and how silicone speaks to itself.

Webster defines technology as "the science of application of knowledge to practical purposes." The Delphic oracle could not have been more cryptic in the use of the terms science, application, knowledge, and practical purposes. One can make a thousand windmills out of them, and they would all spin to the wind that preordains them. At least one interpretation of technology is that of just spelling out a fleshless, knowledgeable, practical, purposeful word or merely the legal existence of one.

If such technology is coherent, its existence proposes a "divination" that is beyond sex, grace, love, and beauty. It is sexist to the point of disposing of sex. It graces itself of disgrace. It loves to confound love. It is the blind beholder that idolizes beauty. But the question here is whether in pursuing the line "the science of application of knowledge to practical purposes" we are losing sight of the time purposes. Supposing we were to substitute the word real in Webster's phrase "practical purposes." We would immediately see that technology is not the end result.

Is it the purpose of life to develop technology so as to produce life larger than life? In biological terms this is just what evolution has been doing for a span of time that has cosmic dimensions—over 3 billion years. With the shift of emphasis from biotechnology to technology and from evolution to metaevolution (evolution conscious of itself), one has to expect all sorts of crises and disruptions. So much of today's turmoil could be optimistically seen as an excruciating invocation and an exorcism of the living upon the nonliving in an effort to make the nonliving into an "image" of the living. We are brutally dedicated to it so much so that we might crush life in a technological embrace.

## COMPLEX VERSUS COMPLICATED

So how do we handle the demon technology? I mean that which we narrowly recognize as technology.

Why is it that so often technology bloodies biotechnology, the living stuff? Quite probably, it is because of the mindlessness of technology. The instruments and the equipment we invent are deprived of mind. They cannot discriminate, or they do so only coarsely. They are mineral stuff only slightly rearranged and reorganized to perform single, mindless tasks. Without a master, they are matter coerced into absurd profiles. A cursory inspection of the junkyard confirms the coarseness of our technological civilization.

But this is not the sin of technology. It is more the infantility of the conceiver and maker (wo-man) who has a hard time staying mentally and emotionally together. One is only embryonically mature. The facts might turn out to be as such only by way of an excruciating working together of "mind and minerals," so that a new level of understanding and creativity will become possible.

In the realm of biophysical stuff the subtleties of the flesh-mind complex go for great complexities in their logistics feedback, interactions, and interdependencies. The flesh-mind system is highly sensitized to the parts that are themselves complete entities. Compared with an elementary biological part, the most sophisticated machines are still machines. They are complicated mechanical-electrical-chemical organizations without being complex. They are at least only mechanical devices without that inner drive that marks even the simplest aspects of the biophysical systems. Because they are complicated constructions, they perform marvels. But they are not conscious of the total system, and they are certainly deaf to their environment and to the mood of their operators/programmers.

I think this is where the danger of technology lies. It takes over where biotechnology leaves off, and as effective as it often is, it is also brutal. It often tends to put things in jeopardy because of a lack of self-control. The examples, such as "leave the stove on and the house might burn down," "leave the centrifuge on and the cream will butter up," "leave the engine on and the train will demolish the station," permeate the entire technological spectrum.

Of course, the robot revolution is based on self-control and self-regulations that imitate the flesh-mind. Irresistible as its appeals are, we are willing to submit and be overtaken (evidence the automobile mystique). The complex (the living) might then be ultimately victimized by the complicated (the nonliving) instead of being sustained and incremented by it. That is a regression, a loss of anima. The consequences are visible in the worst manifestations of Western society. The manifestations can be thrown into a pot, and the pot can be labeled materialism. Materialism is a pathology of technology. More precisely, materialism is the pathology of a society that, unable to discriminate between feasibility and desirability, takes the road of less resistance and drowns life in a sea of innocent, reordained mineral stuff.

But de facto technology, as an extension of another technology—the biotechnology of organic life—is ephemeralization of reality, the process by which a relatively dull universe of mass energy takes time (in space) to recreate and metamorphosize itself in more subtle, sensitized, and ultimately loving occasions.

More and more clearly we are becoming the filters and the retorts of matter. The stuff that we may define as the primeval medium reorganizes itself into tools and machineries. In that sense, technology is setting up a neonature that we in the West maintain to be more direct and more participatory in the venture of "becoming," as in the case of iron ore becoming useful tools. If so, materialism as an inclination toward the enchantment of the manipulation of matter might well stand for spirit at the embryological state—a state that is more advanced than that of matter.

In our everyday dealings, we tend to use technology to simplify our surroundings. In so doing, we tend to simplify reality itself—a process clearly antithetical to the whole direction in which evolution is progressing. Our analyses are essentially processes for the simplification of reality. We have become masters in the art of simplifying whatever gets into our hands; it is the synthesizing that escapes our attention, skills, and capabilities. Technology is the method of our reduction in simplifying the complexities of realities. In so doing, we forget that complexity is tied to values and validity. In most areas of knowledge and life we are still in the analytical stage of knowing and doing, which is evident in our modern life-styles (notably in the areas of medicine and suburbanization). In the name of suburban living, we have split ourselves and isolated ourselves from the very realities of existence on this earth.

It is clear that we in the West are manipulators and transformers of matter that is available to us for manipulation. By reorganizing matter we are forever bringing matter closer to us by way of technology. The transformation of matter into objects has become so paramount an objective that in its pursuit, we have exceeded all reasonables. Producing "something" (or "anything") has in itself become a goal, subservient only to our goal of finding a market for that which we have produced. Our modern shopping centers are good examples of showcases that exhibit the products of our manipulation of matter. These products sit there asking to be bought. Most of the time they are bought and discarded in the same instant.

Individually we are immensely complex bundles of flesh congregating into complex systems—social, economic, and cultural structures, and the institutions contained in them. This congregation is then sustained by complicated servo-systems—the services and the utilities expressing themselves. They are expressed by the complex physical system—the habitat.

## Feasibility and Desirability

The cumulative and self-exacting effects of technology saturation of humans has the mesmerizing and ultimately maiming effect of dispossessing humanity and enthroning the "idol" machine on the vacated seat. This is not happening in the form proposed by horror movies, but in the contextual, ordinary form; it is present and developing in a society at odds with its own affluent, buffeted, insured, and (by all forecasting) happy milieu. Mediocrity of aims automatically produces mediocrity of means. The worst results are the loss of anima, the mediocrity of the spirit, and the mediocrity of meaning within a mediocre ideology. It is natural for a soul slowly losing its own dignity and its own self-respect to turn to physically tangible, reassuring, and laborsaving mechanisms to prop up the weakening structures of reverence and self-responsibility. The consequence is the opposite of the ephemeralization of healthy technology. It is the burial of the dance of life in an equipment yard of obsolete objects robbing humanity of its resourcefulness and true creativity.

Since manipulation of matter is the love affair of humanity with Mother Nature, mainly a male addition (homo faber), we must become more discriminating and orient our passion in the direction of the desirable. That means that our exploration and colonization, which constantly seek new limits of the feasible, must be stringently disciplined. That also means we must break the irresistible propensity that through salesmanship one translates feasibility into marketability. Unfortunately, we are learning the hard way that the feasible is not necessarily the desirable. We are now in the process of selling to outselves the desirability of atomic warfare, the death wish of consumerism brought face to face with its own nemesis.

We truly belong to a monumentally ambiguous project that optimistically is engaging the living in processing the nonliving. Pessimistically, we are engaged in a process that is smothering itself by the onslaught of matter extravagantly fashioned into psuedoorganizations or organisms cluttering a shrinking human landscape. Obscurantism is incipient in our superlighted, electrified habitat. Obscurantism is a dissipative condition, but it is not a dissipative process in the Prigogine grammar; it is entropy-prone without offering the redemptive product of the greater sophistication of mind-spirit. It lacks the discriminatory skills of "light," grace, and aesthetic affirmations.

Discrimination is the educated choice, and as such it is the opposite of segregation. Education is then the priority, so long as it is not merely the present technocratic education that is entirely devoid of all human aspects. Our modern-day technocratic education has become the very thing that is wrong with our educational system. What we need is sensitization of our sensibilities and not just learning by rote or storing irrelevant information for the sake of having information. A fundamental distinction needs to be made here between what is desirable and what is feasible and marketable. In fact, to attain what is desirable is to achieve that which is challenging and perhaps even impossible.

## STEWARDSHIP—TRANSFORMATION

Webster is more helpful in the definition of techne than in his definition of technology. Techne is defined as "the principles or methods employed in making something or attaining an objective." This definition fits hand-in-glove with my proposition that technology is the principle or the method(ology) employed in making the genetic structure of reality and in attaining the objective of realizing equity and beauty for all. The making of such a genetic structure (the cosmogenesis) will take all the time needed for the total metamorphosis of reality, all the space (expanding and contracting) necessary for such metamorphosis to implement itself, and all the mass-energy available to structure and fuel the process. The fully concluded metamorphosis will imply the termination of time (change); the implosion of space into zero space; and the exhaustion of the physical, mass-energy reality. The gene of the cosmos will be a spaceless, timeless, mass-energy-less reality comprehending all the evolution that stands behind it, a resurrection of it all—the Omega Seed of cosmogenesis. When comprehended thus, the assessment of technology becomes critical, especially when we put the question in ontological terms. If wo-man is a phenomenon of relevance within the cosmos, are we to be steward of the earth? Or is our task to transform the earth in the direction of the improbable? Are we to be the tranquil upkeepers of the marvelous ecosystem that is slowly and irresistibly recycling itself? Or are we the troublemakers constantly at odds with it and scornful of its complex ways? Or are we developing the knowledge and the wisdom that will eventually carry the ecosystem (cosmos) into spirit?

The adoption by religions of a model of things that is fundamentally animistic precludes the reality of time. A universe that is vibrant with spirit proposes a circular run, a recycling of matter into that whence it cometh. One can stop the recycling anywhere, for the cycle is still intact and indestructible.

At this juncture, it might be worthwhile to examine whole religious assertions as to the state of the earth at its inception and those revelations relating to the god-created earth. The assumption of the "perfect" in the beginning is purely a religious notion that precludes the opportunity to find, make, and attain perfection. The god-created universe theory proposes a reality that is basically cyclical, where all points are equally valid, and the recycling continues intact even when the cycle stops. Time in this model is reversible and exists in the de facto reality in which consciousness, good, and evil are predefined and predestined. This "frozen perfection" is a cosmic deception perpetrated by many religions. These visions have permeated our world view of what the earth is all about. Technology is being accordingly understood and fashioned. When we understand the whole of human evolution in terms of the "infinite number of presents" making up the history of the cosmos, we tend to understand that our role is merely to maintain the present. This has been the greatest impediment to having a broader technological vision.

The assumption of perfection of the god-created universe is comforting only because it allows us to go on with the fatal model wherein technology is the steward. This may serve the purposes of a religious view of what creation is all about. But it does not address such questions as "What if time is real?" "What if becoming is a creational process?" "What if we are part of the creational process via our intellect?" "What if the cosmos has begun to seek its own genesis—the creation of its own seed for instance?" "What if there has been a real beginning, and therefore, there must lie a real end?" "What if becoming is the needed interval for a dumb media at the beginning to become a pure, radiant, loving, and beautiful message at the end?"

If this were to be so, then technology is the deus ex machina of genesis in progressing from cosmogenesis (the immense mass energy reservoir of the beginning) via biogenesis to the infinite grace of the end. Self-effacement would then become a mandatory parameter of technology. It is then miniaturization, the other side of the medal of complexification conjuring with it the metamorphosis of time into duration. Because it is only with the progressive effacement of time and space (their consummation) that spirit will progressively pervade the real to the "inconceivable" point or moment of pure mind-spirit.

## SUMMARY AND ANTICIPATIONS

What in the short term is the anticipatory scenario of hope for assessing technology?

1. The dropping of feasibility in favor of desirability;
2. A leaner and more equitable predisposition of things;
3. A sharper perception of the nature of reality with the practical ultimately subordinated to the real;
4. The defrocking of intolerance and greed to their clear and coarse pseudotechnological nature;
5. The parallel liberation of the technocratic mind to allow its sharpness to operate at the human local level instead of working at the logic-of-profit level;
6. The perception of how easily deception and self-deception operate at the heart of capitalism as much as they do at the heart of Marxism—materialism is eroding the core of them both;
7. The disconnection of technology from materialism, since via desirability it is technology that can work at the metamorphosis of "matter into spirit";
8. The realization that it is in this power of transformation that technology must not only be explained and justified but also be recognized as the hope of life. Technology is the ship capable of taking life into the ultimate end—the spirit—in the process (evolution), taking with it all of reality.

The practical mind has had it long before reading this chapter. It has had it long before, when for the first time it got reality confounded by practicality. The aim of life is the practicing of reality. The main activity of practicing reality is the creation of reality; the creation of reality is fundamentally foreign to what goes on in the name of being practical in our contemporary world.

Being practical has come to mean conservation and/or conservatism by which our tomorrows are being planned with yesterday's tools and equipment, of course, with the needed adjustments. But reality is a process of transcendence; it is a self-creative proposition endlessly on the move. If the technology does not develop to sustain the proposition, technology is courting entropy. It is ballast instead of fire. Therefore, we have a desperate need for realism so that we can ask from technology the things that will foster reality. Realism implies the recognition of ambiguity, the recognition that as soon as things are very clear the alarm must be struck. A very clear reality is a misnomer: an image might be clear; what it stands for is just about unknowable. Therefore, our quest and our action (the desirable) are fraught in a mystery only slightly dissipated by the fragmentary knowledge of the past and by our "anticipatory realism." With those two flimsy crutches we must do our best to develop the technology of transcendence, the direct descendant of the technology of the flesh via the brain-mind.

So what? While Western civilization endorses metamorphoses, when it actually comes down to it, it acts as if metamorphoses were the edoism of mind. The Eastern mode of civilization is right inasmuch as it scorns the conceptual fuzziness of the Western frenzy for manipulation; it is wrong in dismissing, at least ontologically, the crucial, essential, theological nature of manipulation.

So what then?

Mindless technology—the technology of profit—is symptomatic; it is not causal. To be a technocrat for profit is to be one's own nemesis, and personal nemesis pulls down all those other gullible skeptics, envious, enraged, friend, or foe.

## A PERSONAL RESPONSE: ARCOSANTI

On a personal level, I have been working in the last ten years on a project that attempts to be coherent with the norms suggested by the preceeding discussion. The project is Arcosanti. An urban laboratory (as I can see it) is in the making; it is a prototype for a learning-doing community of about 5,000 persons. The preeminence of technology, in the larger meaning developed above, is critically clear in the premises the project proposes. The urban effect is the drive within the living process—any living process.

The proposed central concept of Arcosanti assumes that the infusion of consciousness within the cosmos is what triggered the beginning—a weak and insecure evolutionary progression—and that, intrinsic to such progression, is the complexification of reality. The enormous complexity of the cosmos eventually pales in comparison to the immense complexity of the mental processes, as well as the mechanisms and the physiological structures that support them. This process of complexification is labeled the urban effect because the city as a system, more than any other, sees the transition of the already unbelievable complexity of the wo-man phenomenon into even more improbable occasions: the political, economic, social, cultural, scientific, and technological totality, which is more than the sum of all its parts and is the epitome of complexity.

The urban effect, then, is proposed as the recurring and imperative effect running through evolution from the very early unicellular organisms onward. It is characterized as an ever-growing force by the implosion of a relatively indifferent milieu, the nonliving universe, into discrete, discriminately complex, and necessarily miniaturized systems. Therefore, the city that promises an access into the future must be that habitat that abides by the imperative of complexity-miniaturization-duration.

Because the urban effect is to be found in every living thing, it transcends each of them. Because we are living things, we find transcendence in the urban effect—that fundamental phenomenon in which two or more particles of physical matter begin to interact in ways other than statistical and fatal (the law of physics).

The urban effect begins in life with not much more than a chemical sympathy between elements and achieves great power in the exploits of love. If the urban effect is a generic concept, it is so because it is a generative and pervasive stress that devours its children in order to reach for more self-knowledge and the integrity implied in its bonding power.

In primitive forms of life, the extension (size) of the associated systems (coral reefs and so on) is determined mainly by environmental factors (moderate climate and so on). The components (the coral polyps) are unconcerned about the extension of the association, because the association is deprived of deep psychoemotional ties and is also logistically loose. If the food is present, the climate favorable, the enemies controlled, and the building blocks within reach, that is all that is needed. The more we move toward complexity, the more the extension of the associative systems (bees, ants, humans) is determined by inner factors, because those inner factors assume greater and greater significance and power.

The coherence of the process through which the cell becomes the city is based on the exponential assertion of complexity by which information becomes knowledge and knowledge becomes the new mover for creation. This coherence cannot weaken until the ultimate condition is achieved.

# PART II

## SELECTED COUNTRY EXPERIENCES

# 4

# ENERGY TECHNOLOGY ASSESSMENT AND SOCIAL AND ECONOMIC RECONSTRUCTION IN INDIA: A TEST CASE

MANGALAM SRINIVASAN

## A THEORETICAL FRAMEWORK

The perception that technology is a force of economic and social reconstruction whose currents set in motion the shape of developments to occur is widespread. The marvels of technology and the eagerness to acquire it have tended to dampen our awareness that technology is relational knowledge that helps in mediating between what is now and what our visions want to achieve for us. Technology is both the tool and the fulcrum on which the lever rests and by which paradigm shifts are negotiated. The general assumptions underlying the process of energy technology assessment often are centered in the concerns relating to the locational aspects of the energy resources and the technologies that have to do with its recovery and use. In the plenary sense, while this is indeed true, the particular aspects of cultures, world views of the communities present, and their philosophies of life are all component technologies in energy use.[1]

## ANALYSIS, DATA, AND INFORMATION

There is an increasing consensus regarding the place of analytical imperatives in decision and policy processes; the difference of opinion seems to center around the expression "analysis at what cost?". The traditional cost-benefit analyses are usually budgeted for in projects; but instituting procedures and programs to assess the full range of consequences resulting from the introduction of technology presents budgetary, time, and personnel problems.

Information is vital to the integrity of the whole policy process. But information alone will accomplish little as has been demonstrated by accurately written trivia flooding the development arena. It is the approach to knowledge and not information that highlights the instruments of communication and the critical mass of concern for a workable and equitable energy policy. In assessing energy technologies, therefore, it becomes imperative to evaluate the whole process, as well as the particular technology under consideration. Although data collection is an essential imperative for any energy analysis, good data base alone does not guarantee a good energy policy, especially when how the problem is perceived, how priorities are assigned, and what urgency is attached to the resolution of the problems are themselves issues of major concern in national energy policy.

## CONCEPTUAL APPROACHES TO ENERGY ASSESSMENT

A rigorous system of energy technology assessment offers a good agenda by which energy decisions are improved. While at the project phase a limited agenda is almost always better than a large and complicated one, at the assessment phase better evaluation of the issues, alternatives, and options requires an exhaustive agenda. Many developing countries have highly complicated and stratified societies whose traditions and belief structures go back to distant times and carry accumulated knowledge of their environment. Each society has its own definitions as to what are tools and ideas and how best to use them. These dimensions add additional variables to the assessment equation. All these factors inhibit the use of strict methodological approaches and statistical truths in energy analysis in the context of developing countries.

The first stage of analysis leading up to the assessment is necessarily ordinate and concerns itself with the following steps:

Locating sources of know-how regarding the energy technology being assessed; establishing a list of who-knows-what regarding the area, subject, process, technology, and product of assessment;

A preliminary, ad hoc searching for technological options;

An initial listing of all costs and all benefits;

A listing of experiences (both beneficial and adverse) regarding the introduction of similar technologies to similar societies or locations (essentially didactic);

Identifying areas of cooperation between the government-private sector, on the one hand, and the community of users, on the other, and mapping out participatory stages;

Identifying particular energy-need-search, expectation, confidence, and fear factors;

Mapping out the linkages between energy policy and other variables in development planning.

These steps basically spell out the assessment parameters in the broadest possible terms, setting the stage for a focused, location-specific evaluation of the technology in the context of a given community's social, cultural, and economic characteristics.

The second stage of analysis is substantive in nature. The analysis consists of the following steps:

Specific features of the framework are described: for example, scale-specific, geography-specific, culture-specific, and quantity-specific categories that narrow the scope of analysis and allow it to concentrate on a single theory or a practice at a given time. In a vastly heterogeneous country like India, for instance, national policy is typically a sympodium of numerous regional and local action measures and resolutions. Where this has not been the case, as in the national energy policy process, there has been a severe systems failure, especially in rural energy planning.

Specific features relating to the methods of energy analysis are the following: what can be added to the information (interpretation), the various connections (linkages), knowledge and investigation of the problem at hand (internal expertise), and the priorities and choices (in-house interests and biases).

Specific features relating to the methods of energy administration are identified and detailed: methods of energy administration are the specific political technologies that advance the cause of policy. The assessment of these technologies reveals what is feasible from the political-institutional angle, what planning should go on, or what help could be provided by government with regard to advancing the cause of creative living and leisure activities and improving the quality of lives of the people in anticipation of the time that would be saved by the introduction of energy-efficient and timesaving energy devices. In this stage of analysis are included considerations related to the size of the systems; the exact nature of the energy conversion devices; the social responses to economic consequences of the system; costs, inflation, and exchange factors; and finally, the institutional aspects.

The third stage is synergistic in that it brings together the information, analyses, and insights from stages one and two. The background information and analyses allow the policy planner to classify, tabulate, weight, and rank alternative strategies, options, and policies. It is at this stage of assessment that opportunities exist to integrate the particular consideration in the overall context of energy planning. The process consists of deciding on the appropriate policy instruments and establishing adequate institutional mechanisms to mobilize human, technical, and financial resources.

Technology assessment does not replace the decision-making processes but is an important aspect of them. It is a critical process by which a full range of future consequences owing to a technological choice are determined before the impact. For this reason, it is imperative to arrive at a theoretical understanding of what energy technology assessment is all about. This forces the analyst to examine the basis from which decisions are being made. Such an understanding will enable the analyst to view technology assessment in a continuum during the different stages of the technology choice, transfer, introduction, development, and use; at completion the assessment still continues in its attempt at evaluating the results against the goals as originally set and in terms of its costs and benefits. A well-organized assessment process not only is a guide to policy, decision, and action but is a frame of reference against which results are evaluated.

In the search for a schematic expression that would best describe technology assessment, this study has taken help from botany. In botany sympodium is a stem composed of a series of branches superposed in such a way as to imitate a simple axis, as in the grapevine.[2] Sympodium at once stands for linear dimensions of length and breadth, the added dimension of depth, and a fourth dimension superposed with interlinks and intralinks establishing a spatial connection among all dimensions present. When this sympodium is placed in a time frame (the fifth and one of the most important dimensions), what we have is a model of technology assessment whose boundaries allow for a comprehensive and useful analysis resulting in the establishment of a workable agenda and valid priorities.

The sympodium concept illustrates how issues and concerns are intertwined and superposed and how the analysis of one almost always involves the analysis of the other components. The concept is especially useful in energy-related technology assessment, because energy is a common component in all activities. At every stage of energy technology assessment it will be necessary to invoke value-oriented and other nonquantitative criteria and integrate them into the analytical process. For instance, in introducing solar energy technologies it is necessary to consider all interacting aspects, such as innovation; the manner in which they are introduced; and the choice, transfer, installation, maintenance, repair, and use. Each stage is a separate microcosm complete with its own set of technoeconomic and social considerations and impacts.

## ASSESSMENT OF TRANSFER AND INTENDED APPLICATIONS

### Solar Energy Technologies in Rural India: A Test Case

India lies entirely within the Tropic of Cancer and as such receives plentiful solar radiation. With an average solar insolation of around 2,500 kilowatt-hours per square meter a year, the country is a prime candidate for large-scale solar energy research and development undertakings. Solar energy technologies are nothing new to a tropical country, where people have traditionally depended on this source of energy for accomplishing a variety of everyday and special tasks for centuries. Although the sun shines most bountifully across the subcontinent, the tremendous diversities obtained in the country do not lend themselves to a standard single solution for the Indian energy problem. This makes energy technology assessment difficult, especially where it concerns the rural energy situation. The vast heterogeneity of rural India inhibits sweeping generalizations and uniform methods of assessment. Not only do great differences exist in resources and problems, but they also exist in need perceptions and cultural factors affecting the choice of technologies. These differences notwithstanding, the government of India must come up with policies based on some sort of generalizations and common denominators. Energy technology assessment by establishing the correct focus and by offering guidelines can help in problem assessments resulting in the establishment of valid priorities.

In 1974 the government of India announced the formation of a committee to study prospects for the utilization of solar energy. The panel is overseen by the National Commission on Science and Technology, indicating the government's willingness to develop the technologies connected with the use of solar energy.[3] Despite the commitment to the development of this energy source, there seems to be a waste of resources among too many marginal efforts spread among multitudes of agencies. Because of the lack of large-scale pilot projects supported by policy, commercial interests are not too eager to enter the field. A number of leading institutions of research in the country, such as the Bharat Heavy Electricals, the Solid State Physics Laboratory, the various Indian Institutes of Technologies, the Central Electronics Research Institute in Pilani, and the Tata Institute of Fundamental Research are all involved in the research and development of solar energy. Again the research efforts are too thinly spread as are the resources. It is being pointed out that policy has been responsible for the following inadequacies and shortcomings of the Indian solar energy program.

1. The goal as set by policy in developing the solar program is to produce a 100-watt panel by 1981. Given the dimensions of the energy problem and the size of the constituency, the goal is too modest.

2. Policy seems to favor all approaches equally. This position shows a lack of deep study, a lack of sensitivity to the priority energy needs of the villages, and a lack of direction by policy in setting up specific projects.

3. By spreading resources too thinly among various competing interests, not only has cost-effectiveness been compromised but a serious time lag has been created, delaying development efforts.

## Solar Energy and Rural Energy Needs

In the rural energy scene the following assessment of solar energy possibilities are the basis for priorities:

1. Household energy makes up a large part of overall energy use in developing countries. In rural India nearly 58 percent of all energy use is in the domestic area.[4] How does the intended energy technology application address this problem of fuel use in the domestic sector?

2. How does the introduction of the new energy source or technology affect the existing energy system? In other words, how does the introduction of, say, a new solar device, such as the solar cooker, affect or impact upon the use of wood fuel and human energy supply (and, therefore, employment)?

3. What are the impacts owing to the introduction of labor-saving energy improvements on different sectors in alleviating human drudgery? For instance, at the present time women spend considerable time in collecting, sorting, and processing fuels. Most of these fuels burn inefficiently and constrain the time budget considerably, especially in rural areas.

4. Who are the primary beneficiaries? Is solar energy equitable to all the people? For example, rural electrification schemes have hardly benefited the domestic sector or any other nonagricultural sector. Rural electrification has primarily meant benefits to irrigation-related activities. In fact, most rural communities have been bypassed by most conventional energy development programs.

5. There is a geopolitical nature to the source of energy under consideration for use in rural areas.

6. The matching of energy devices to the energy resources locally available and matching them with the felt needs of the people are necessary. For instance, using solar energy for space heating in much of the tropics is a wasteful activity, whereas harnessing it

through technology for domestic cooking and lighting uses, crop and harvest drying, dehydrating foods, and so on contributes to vital activities in the lives of the people of India.

7. Can community energy systems be strengthened by the introduction of the new solar technology, such as for space lighting and heating in community halls and for use in community kitchens, schools, health centers, and so on?

Sophisticated comparisons involving irrelevant and complicated categories are not useful for determining what will work for the rural people. For comparisons to be useful, they have to include the cheapest available alternatives. A majority of the rural inhabitants are too poor to afford anything that will require even a marginal addition in their low-level stagnant finances. The assessment of any device, however attractive, will have to stack up reasonably comfortably against the existing energy device (mostly based on the so-called free fuels[5]) and should not add significantly to the time budget. Given the significant amount of time spent by rural families in collecting fuels, the devices being assessed should hold promises of saving time, as well as energy.[6]

Comparisons of alternatives will involve issues of broader significance both to the local community and to the nation in terms of environmental and ecological impacts and balances. While some innovations may turn out to be cost-effective, they could deplete the resources to an extent that such technologies and alternatives cannot be recommended even if the cost approaches zero. For instance, introducing an efficient tool that may harvest wood fuel in large amounts in a short time is clearly to be evaluated against the available wood resource, the growing cycle of the species of wood it will harvest, the other uses of the trees, the effect of large-scale cutting of trees on rain and soil erosion, and what plans there are for the continued growth of the wood species so that at all times the optimum quantity of growing trees is maintained.

The need for laborsaving devices in the rural areas where muscular energy dominates the entire spectrum of rural living and has kept most of humanity living in these areas bound to drudge labor.[7] Policy rarely concerns itself with the drudgery that is a large part of rural work, be it in agriculture, crafts, or rural households. It is customary to evaluate the project or a particular technology in terms of the contribution it makes to the gross national product and national development goals of income and productivity and not in terms of its impact in reducing drudge labor for people in the villages.

There is little doubt that more purposeful studies are needed before any new innovation is introduced for use by people in rural

India. It is necessary in the case of solar energy uses to know as accurately as possible how solar technology is being introduced, who is introducing what technologies, and in what manner. The concepts relating to the use of solar energy are well understood and accepted by the rural people; it is the new technological dimensions and their external controls that they have to cope with. The traditional technologies in the rural areas have always been in the hands of the people; the modern ones are controlled by technocrats in the government from centralized locations. In evaluating the introduction of solar technologies, the following points will have to be considered:

Who is doing the introducing—local leaders, government representatives, technicians, or foreign consultants;

How dependable the visits are of those who are charged with the duty to explain, teach, and train the local people in the use of the device;

How theories and practices concerning the device are being introduced; whether they are introduced in well-ordered, learnable stages or whether the whole system is being delivered in packages with instruction manuals without relation to training and imparting the necessary skills.[8]

Scale of technology is an important consideration in rural India, where size can cause serious disruptions in the life-styles of the village people, as well as in the rural ecology. Some size decisions also bring in sociological considerations, as in the cases of community kitchens and community forestry where, owing to existing social and economic stratifications, the projects have resulted in failures. Economic considerations clearly point out that community systems are preferred over individual or family uses of the devices to be introduced. This is where technology assessment will have to make the difference—by supplying dependable information as to why proper scale is a necessary part of the decision and what policy instruments may be needed for deciding upon the desired goals in the interest of rural development.

Management of the solar devices installed ultimately will be the major factor in deciding the continued success of the results of the technology. Time and again development experiences have revealed how useful technologies have been rendered totally useless because of mismanagement. Lack of local participation has also led to serious impediments to program implementation and project benefits. Perfectly designed and superbly executed technologies are doomed to fail if they are not integrated into the village society. The management required differs not only from one technology to another but from one culture to another and from one community to another. The type of management that will succeed is a resultant function of the local interest; local participation; proper dissemination of information regarding all aspects of technology; profitability of the project; and proper distribution of services and benefits of the solar technologies to the intended sectors, groups, and individuals.

In the Indian context, it would mean investigating the village panchayats, or councils; the local public works departments; state or central government representatives' offices; and other administrative infrastructures in terms of their technical and managerial capabilities, staff support, and attitudes. The colonial memories are still alive in the rural areas, and the Indian bureaucracy has done precious little to alleviate the misapprehensions associated with the government. Mutual distrust persists between the government administrators and the village communities. The villagers are more often than not intimidated by the administrative styles, while the administrators have been unwilling to have the necessary confidence in the ability of the villagers to learn, manage, and succeed in their efforts. All these elements need to be looked into and properly recorded for the benefit of policy makers so that they can develop the appropriate managerial-administrative infrastructures prior to the inception of the program.

## Technical Considerations

The integrity of the whole system depends on proper technical considerations. Design considerations include, apart from the correctness of the technicality, the factors that determine the choice of materials, the thermal and mechanical strength of properties, the processes and methods by which materials chosen are assembled, the mechanical climate controls, and the analytical and experimental features. For instance, the passive thermal system, such as the simple parabolic mirror solar cooker[9], should describe all the construction stages, such as determining the parabola and making the model, and then from the model constructing the cooker and coming up with operational and maintenance procedures and instructions for its continued performance. In the case of a solar room heater, technical consideration will involve determining the panel size, the heat-storage media, and mediative and convective heat losses from the surface of the panel and from the room air. As for other solar devices, technical considerations may consist of materials, construction details, cooking efficiencies, local manufacturing capabilities, adaptability to local cooking requirements, durability, and so on. Technical assessment is not independent of locational, sociocultural, and other specificities affecting each given use of the technology being considered.

## A Framework for Comparing the Alternatives

Assuming that the rural consumers are not currently in a position to stack and rank the technologies to be introduced and are not directly involved in product choices or policies, it would be necessary for the technology assessor to study in detail all relevant choices in terms of costs, efficiencies, performances, durabilities, technological simplicities, and so on and list them in a table assigning numerical values to each of these factors to help guide policy. For example, the numerical values of several solar cookers to be introduced in selected rural villages of India may be devised by using Table 4.1. The solar cookers are then evaluated against the criteria listed with an assigned numerical value of 1 to 5 where the higher number represents the greater promise in terms of its potential usefulness to the target community.

TABLE 4.1

Numerical Values of Solar Cookers

| | Type | | | |
|---|---|---|---|---|
| Criteria | A | B | C | D |
| Design aspects: suitability for local cooking of traditional foods | | | | |
| Cooking performance | | | | |
| Efficiency | | | | |
| Cost | | | | |
| Are tools and materials needed for construction locally available? | | | | |
| Simplicity | | | | |
| Possibilities of acceptance | | | | |
| Dependability | | | | |
| Future improvements | | | | |

## Technology Evaluation

### General Considerations Relating to Resources

The level of knowledge of the resource base is an important consideration "in the mobilization of the resources as a continuous link from production through conversion to end-use as a unified system."[10] Knowledge of the resource base needs to be extensive not only with respect to the present supply but also in terms of its future prospects. At its present stage of technology and cost, many solar technologies currently off the drawing board probably cannot be introduced into areas deficient or sporadic in insolation. A complete data base listing average annual insolation, number of cloudy days in a year, current patterns of use of this source, and so on is to be constructed prior to project siting considerations.

### Geographical Factors

Average annual solar radiation varies widely within the subcontinent and is influenced by specific locational conditions, such as the proximity to the Central Indian Desert, the sea, the rain forests, the mountain ranges, the cloud cover, the height differential from the sea level, and even barren flatlands.

It is evident from what available information there is that the dependability of solar radiation for a particular spot at constant intervals is of paramount importance for this source of energy to make a meaningful difference. The experiences of other countries that have gone ahead with solar experiments have also been the same. For example, the southern part of Taiwan is located in what is called the maximum solar energy band (Copper 1976, p. 257) and receives more intense sunlight and relatively less cloud cover than other parts of Taiwan. The country has several solar water heaters and other solar devices in operation both in its domestic and in its industrial sectors. In Taipei, for example, there are a large number of solar water heaters, solar space-heating systems, air conditioners, and refrigerators in operation. But the operation is being constantly interrupted by the many cloudy days that the city has and the lack of storage mechanisms. Added to the above are the large considerations inherent in the installation of solar devices in their present stage of technology, namely, the technological maturity and the cost of this energy source.

### Technology Feasibility

Even the most promising solar technologies are found to be inadequate in field testing and siting owing to any number of reasons, the most important of which concerns the maturity of the technology under consideration and the high cost of installing it. Solar technologies are not yet viable for application in rural areas without massive government subsidies. Private investments are not readily forthcoming owing to the lack of large-scale political commitment and profit possibilities.

## Scale of Operation

Solar technologies offer the best possibilities yet for decentralized power self-sufficiencies (especially in rural India) if and when these technologies become cost-effective and local in resource base, production, and distribution. Photovoltaic cells and solar collectors are technologies that are well adapted to dispersed siting.

## Demand Management

One of the reasons for the failure of many new technologies, especially in rural areas, has been the inability of public administration to match local demand with the choice of technology and the product of such technology. Efficient end-use of energy is an important aspect of total energy management, and as such the assessment of whatever solar technologies are under consideration must include how and in what manner the efficiency of energy end-use may be improved.

## Assessment of Assumptions

Since analyses are often predicated upon assumptions that have been made, it would be wise to state all assumptions and evaluate each of these assumptions for its validity. Assumptions may be technical, economic, sociopolitical, or about climate, environment, or ecology. They will have to be carefully analyzed and stated preceding each alternative under comparison.

## Operational Problems

It is clear that when a perfect system has been put in place it can be ruined by any number of factors, such as the lack of trained personnel who know how to operate the system with advantage, faulty procedures of maintenance, nonavailability of proper spare parts, and inability to repair a systems breakdown. In addition, the intensity of solar radiation may directly impede using solar devices. For example, Table 4.2 (Parikh 1978) amply illustrates the relationship

TABLE 4.2

Monthly Variations of Solar Radiation and Typical Efficiencies in Two Cities of India

| Month | Days | Nagpur | | | Jodhpur | | |
|---|---|---|---|---|---|---|---|
| | | Average Radiation (calories per square centimeter per day) | Utilization Efficiency | Monthly Availability (calories per square centimeter) | Average Radiation (calories per square centimeter per day) | Utilization Efficiency | Monthly Availability (calories per square centimeter) |
| January | 31 | 460 | 88 | 12,549 | 410 | 84 | 10,676 |
| February | 28 | 510 | 83 | 11,852 | 480 | 84 | 11,290 |
| March | 31 | 570 | 76 | 13,429 | 560 | 77 | 13,367 |
| April | 30 | 610 | 72 | 13,176 | 630 | 78 | 14,742 |
| May | 31 | 630 | 70 | 13,671 | 680 | 83 | 17,496 |
| June | 30 | 480 | 40 | 5,760 | 680 | 73 | 14,892 |
| July | 31 | 400 | 23 | 2,852 | 530 | 48 | 7,886 |
| August | 31 | 380 | 23 | 2,709 | 480 | 43 | 6,398 |
| September | 30 | 480 | 51 | 7,344 | 540 | 54 | 8,748 |
| October | 31 | 500 | 69 | 10,695 | 510 | 84 | 13,280 |
| November | 30 | 480 | 87 | 12,528 | 440 | 91 | 12,012 |
| December | 31 | 420 | 84 | 10,937 | 380 | 88 | 10,366 |
| Total | 365 | | | 117,502 | | | 141,155 |

between utilization efficiency, the intensity of solar radiation and utilizable solar energy, the monthly variation of solar radiation, and typical efficiencies in two cities of India, Nagpur and Jodhpur, which are about 500 miles apart.

Given the distances obtained in India, it would be rather impossible to devise a policy that is applicable in all places and under all conditions. Operational difficulties are encountered if the procedures are not location specific. Parikh (1978) points out the effect of the variance on the performance of solar pumps in the following:

> One can see that in Nagpur, the utilizable energy drops by a factor of 5 between the months May and August. In fact, these are the months when water is required for cultivation. The reason why the solar radiation drops is that it rains in this period. In case the rain gets delayed and it is nonetheless cloudy, then the solar pump installed may not be useful, unless the collectors also collect diffused radiation and desired water is provided. Thus the availability of water at a critical period, which determines the success of a crop, is essential.

## Social Assessments of Solar Technologies

Philip Palmedo defines energy technology assessment as follows:

> Energy technology assessment is the systematic evaluation of the implication of alternative technological means to fulfill society's need for energy. Assessment generally assumes, implicitly or explicitly, a decision to be made. Most crucial energy decisions are not exclusively technical, and energy technology assessment is normally embedded in a larger policy framework. Neglect of that broader framework can lead to irrelevance of the assessment or, what is worse, the misapplication of its results leading to socially undesired consequences.[11]

Perhaps the most important aspect of technology assessment that guides policy is its readiness to evaluate technology against society's needs, society's requirements, and the existing sociocultural background. In many instances, the hasty introduction of technologies based on partial or unsound analyses has led to serious disruptions to the social fiber,[12] has worsened the economic conditions for the great masses of people, and has caused environmental degradation. Solar applications are extremely attractive propositions for rural India, which has been bypassed by most central energy planning. The decentralized nature of solar applications makes it extremely viable and useful for rural purposes. The use of solar power for agricultural production, community energy systems, and in the rural domestic sector requires that technology assessment correctly assess the agricultural and rural tasks of men and women and, in the domestic sector, the work of women and the contribution made by them in the rural economy. This will enable the matching of tasks appropriately with the solar devices that may achieve the tasks for them.

The existing directory of development experiences is a tale of incomplete assessments of rural-user requirements. For this reason, for each solar application to be introduced and for each sector the following criteria will have to be comprehensively and systematically organized:

1. The effect of the application on the quantum of rural work in general, and within it, the division of labor (how it is divided between men and women and among the various village strata, and so on).

2. The effect of the application on the income structure. When drudge labor is reduced, how can the surplus time be used to increase rural income? How does the application affect the incomes of the various members of the rural and village communities, such as the landed, agricultural laborers, village craft persons, household workers, men, women, and children?

3. Rural areas are currently delegated to the "stagnant microsystem"[13] of subsistence production and existence. What is the effect of the proposed application on this system and the low level of productivity that the system exemplifies?

4. What sociological changes may result from the application of the chosen technology on rural inhabitants? How do applications, such as street lighting, affect the socializing habits of the people?

Many studies have shown that modern technologies have adversely affected the role women play in a rural work force. Automation and other technical aspects relieve women from the main-line duties and relegate them to the first and last stages of rural work, such as preparing the feedstock for the machines and packaging the product while men operate the machines. These tendencies will have to be investigated in relation to the new application being introduced, whether opportunities exist for making the system more equitable among the various rural sectors, and what overlaps may exist between the particular effect of this technological application and those that already are in productive use in village communities.

## Energy Technology Assessment and Public Policy

The Indian public policy is basically the child of British urban thought and, in its present form, is least connected to the needs of the rural people. National planning, for instance, is concerned more about developing a particular resource rather than with the local uses of the resource by rural people, as in the case of rural electrification. This inadequacy needs to be corrected if solar energy consumption through new technologies is to be placed on a fair and equitable basis among the various sectors of the country.

The policy to integrate the rural population in energy development and use is part of the same process of the overall policy to integrate national energy plan targets with local energy objectives and with the economic and social development of the country as a whole. This will involve examining each rural energy context by resources (human and material), the means by which the resources are being developed and utilized, and the effect of the various developments on human and environmental ecology in relation to the different rural sectors.

There are specific inadequacies that are methodological and statistical in nature concerning the data, surveys, conventions, and indexes on rural life, people, and their role in energy consumption and production. Insufficient and inconsistent data have distorted both the nature and the dimensions of the problem as well as the solutions extended. Technology assessment will have to itemize these inaccuracies in energy surveys if these surveys are to help policy. The most common of these inconsistencies to watch for are the following:

Methodological: Surveys are based on methods that have been successful in Western industrialized countries where data are available; where energy materials, use, and practices are far more standardized.

Statistical: The data concerning the rural areas, especially in the energy fields, are, as a whole, not integrated into the overall data picture for the country.

Energy Indexes: Many of the energy indexes used in evaluating rural energy use are unnecessarily obscure and complex. In stating efficiencies, often vastly different items are being compared. Other indexes, such as the heat content, thermodynamic availability, price factors, fuel substitutions, and fuel specificities, do not include the usefulness or properties of widely differing fuels or the values and preferences of the users, especially those of women, who are substantial energy users in the rural areas.

Conventions: The inappropriate or insensitive treatment of the measurement conventions (such as energy embodied in materials, the treatment of animate energy [human and animal power], the distinctions as to the net and gross caloric values of combustible fuels, and the fuel cycle measurements of energy flows) seriously affects the determination of energy needs, uses, preferences, and consequently, the public policy allocations of energy allowances, projects, and programs for the various sectors of the country, including women.

The assessment of solar energy technologies for rural India requires the assessment of specific policies related to the broad categories, such as the following:

Centralized versus decentralized decision units
Land reform policies that affect land tenure and ownership
Administrative coverage and accessibility
Resource control and utilization patterns
Technological complexities
Program linkages and flexibilities
Prices and profitability
Distribution of benefits

In the above category of considerations, program linkages are often left out. Women, for instance, play diverse roles and all at the same time in many areas of rural life. Agricultural programs to benefit women would have yielded better results if they were connected to childbearing and family well-being. In a larger sense, those programs that have been successful are the ones where the rural community itself has been integrated into all aspects of a given project.

## SUMMARY

The theoretical framework put forth earlier in the discussion and the preceding discussion on solar energy technology assessment enable us to organize the cardinal points in the following fashion:

I. Ordinate stage

Consideration: General energy issues affecting rural societies

A. General energy scene
   1. Rural energy endowment
   2. Present energy needs
   3. Future prospects
   4. Options
B. Development linkages: energy and rural development
C. Factors affecting technology demand, innovation, and introduction

D. Policy analysis
   1. National
   2. Local
E. Regional criteria
   1. Ecological
   2. Resource base
   3. Environmental
F. Alternative technologies
   1. Efficient stoves
   2. Labor/time saving devices
   3. Income-producing innovations
G. Energy from biomass and solar energy
H. Data base and methodology
   1. Level of aggregation
   2. Energy indexing
   3. Social science approaches
I. Problems of institutional structures in energy decisions

II. Substantive stage

Consideration: Selected solar energy technologies and rural energy needs

A. Complete review and examination of the existing arguments and assumptions governing rural energy uses, sources, technologies, and strategies and the role of rural people in all of these
B. An empirical analysis of demand and supply aspects of these energy applications and their societal impacts as they concern rural people and their life-styles
C. Economic and social analysis of the effects of energy technologies in the rural areas and their viability in relation to the rural family structure and whether the fuel substitutions are compatible to the type and quantity of cooking and heating activities undertaken by women; many of these new introductions are quite arbitrary as they are not preceded by women's participation
D. An inventory of the types of energy sources, technologies, and applications that have been introduced in the last few years in selected rural areas and how they have impacted on the various sectors in representative countries, which will allow us to draw some general conclusions

III. Synergistic stage

Consideration: Integrating the introduction of the new solar energy technology with the issues of human development in rural India

A. An intimate knowledge and understanding of the social structures, cultural preferences, and skills of the people
B. An efficient organization that can bring the people together, and which is representative of the groups and sectors whose interests it wishes to serve
C. A system of management that will be goal oriented and will optimize efficiency and results, and which has a learned formula for the equitable distribution of benefits; one whose management philosophies are concerned about not only the size but also the shape of the policy measures undertaken
D. A dependable local resource endowment that is cost-effective to use, and which is renewable
E. A technology that is structurally simple, adaptive to local conditions, free of external dependence, and does not replace the local solutions

Based on the above, policy should establish procedures on

1. How to assess the kind of data required and how to obtain them;
2. How to resolve the problems and obstacles to better utilize the existing and new energy sources for enhancing the quality of life in rural areas;
3. What actions might be necessary to enlist local participation in energy decisions, to attract public and private funds to the rural areas to be used for energy resources development, and to improve their technologies for better and more efficient use by rural people;
4. Developing infrastructures that might make the contribution and participation of local people in the energy field economically and socially viable;
5. What training and skills need to be imparted to the various rural sectors so that they could make a more effective contribution toward the development of energy systems and technologies for their areas and could have greater creative participation in all energy decisions that concern them;
6. Corrective measures that may be needed to alter the existing conditions that are inherently discriminatory toward certain sectors, and which may bring about equitable distribution of the benefits owing to the new energy developments.

## CONCLUSION

The nature of rural ecology and society demands a holistic outlook on the part of the analysts. Rural India is what Revelle would call a "partially closed eco-system"[14] in which all activities are internally generated from the given resource base, each with its own indigenous pattern of growth, development, decay, and rebirth. The process goes on in an endless cyclical fashion. When this process lacks revitalization from new inputs and information, the cyclical process rejuvenates itself, only marginally losing a $\Delta x$ (where $\Delta$ represents the rate of change and x could be any variable, such as soil fertility, forest resource, water table, cattle, or any other) at each stage of rotation. This is quite pronounced in the Indian countryside, where centuries of working the soil without adequately reconstituting it, burning forest fuels without safeguarding their continued growth, and working the human muscle without improving the quality of the caloric intake have all but brought the closed ecosystem to the brink of disintegration. It is clear, given the nature of this closed ecosystem, that changes and improvements cannot be made in one part of it without the entire system being subject to some modifications. Technologies introduced casually in these circumstances can turn out to be technologies that can alter the delicate balance of rural survival, disturb the social fiber of the rural society, and not fulfill the demand for basic goods and services in rural India. Therefore, the first step in energy technology assessment is the requirement analysis that determines considerations, such as whether there is a need for the product and whether the product should be locally manufactured or imported. The second phase is documentation relating to the where, when, what, and how of the technology, as well as how to establish markets so that consumers have an outlet from which to choose the product. At the implementation phase, the assessment is mostly concerned with the overall hardware and the operational problems that may be anticipated in the course of the installation and beyond. The next and most neglected phase is what I call the breakdown phase, when problems begin to appear in the operation, maintenance, product quality, and standard connected with the end use. Technology assessment will have to assess the possibilities of the breakdown phase at the stage of the choice of technology itself. Finally, technology is not mere machinery; technology is machine plus method, where the method may stand for anything from procedures, rules, and political/administrative measures to consciousness and intuition.

Solar energy has gained the same sort of respect among progressive persons everywhere. Despite this reverence and a general agreement as to the desirability of this source of energy, especially for the billions in the Southern Hemisphere, at the present time in India it is still largely intuitive and instinctive rather than a policy determination. In the ultimate analysis, only political commitment and appropriate public policy measures and instruments can make this source viable for use in the vast expanses of rural India.

NOTES

1. Technology here is deliberately used to mean not only the hardware or the "machine" aspects of advancing a given scientific principle, but also the software aspects encompassing a whole range of subjects, such as approaches, methods, attitudes, and knowledge used in achieving the desired goals.

2. Webster's Universal Dictionary (New York: World Syndicate, 1936), p. 1689.

3. The minister for science and technology, C. P. N. Singh, during an interview for this study, stated that solar energy is among the top priority energy sources under consideration for research, development, and use, and that (as a scientist) he was leading the efforts toward the best utilization of this source of energy for India's development.

4. Vikram L. Dalal, "Energy Strategies for India: A Technological Perspective" (Paper prepared at the Center for Environmental Studies, Princeton University, Princeton, N.J., 1977), p. 4.

5. In the rural areas of India, except for kerosene for domestic lighting and electricity for irrigation of some of the rural land, much of the energy is derived from traditional and noncommercial sources of fuels, mostly consisting of cattle dung, vegetable wastes, forest debris, crop residues, and firewood. Rural people spend considerable time combing the earth for these "free fuels." Apart from the fact that many of these fuels are entering the markets because of their dwindling supply, there are huge costs involved for the nation as a whole in economic, social, and ecological terms in the form of loss of organically rich cattle dung fertilizers to the land and depletion of forest resources—all leading to environmental imbalance and degradation. Many of these "free fuels" also burn inefficiently and cause considerable ill-effects on the health of rural women, who are the primary users of these fuels.

6. Revelle estimates that in rural India "cooking, obtaining water, collecting fuel and other domestic activities take up nearly 200 billion hours." These details illustrate the magnitude of human energy in use in the rural areas; they also tell us the significance of women's contributions in the whole rural energy field. Roger Revelle, "Energy Use in Rural India," Science 192 (June 4, 1976): 969.

7. It is significant here to quote the following passage from the U.S. National Academy of Sciences study "Energy for Rural Development" (Report of the Ad Hoc Panel, Board of Science and Technology, Washington, D.C.), introduction.

> Energy plays a larger role in man's struggle with the vagaries of nature than merely sustaining life. In The

Economic History of World Population, Cipolla has said that the more successfully man can use his own energy output to control and put to use other forms of energy, "the more he acquires control over his environment and achieves goals other than those strictly related to animal existence." He then adds what is certainly obvious but does not suffer from repetition—that fundamental to the utilization of non-muscular energy is the problem of transforming it into the needed form "at a selected time and place and at convenient cost."

8. Daniel Benore, an agricultural extension services consultant to the World Bank who is credited with massive achievements in agricultural development in Israel, Turkey, India, and a number of other countries, has pointed out both through written and oral communications that the fundamental ingredients of a successful extension program are its ability to demonstrate dependability, trustworthiness, and credibility of the persons charged with the responsibility of the program. He points out that often policy planners do not keep their appointments, schedules, or promises because of the degree of importance they feel about themselves. Benore points out that where it has been demonstrated that the government planners mean genuinely to help the people and that they are serious about such help, the response has been overwhelming on the part of the rural people. Long years of colonial rule and neglect of the rural population by officials have led to the rural people viewing the government with distrust. Therefore, any model constructed to help must include the consideration of how to infuse trust in the policies and programs of the government.

9. M. Von Oppen, "The Sun Basket," Appropriate Technology 4 (November 1977): 8-10.

10. United Nations General Assembly, "Report of the Synthesis Group," Preparatory Committee for the United Nations Conference on New and Renewable Sources of Energy. Third Session, March 30-April 17, 1981, Item 2 of the provision agenda.

11. Philip F. Palmedo, "Energy Technology Assessment: Consideration of Geographical Scale," Journal of Energy and Development 2 (Spring 1977): 207.

12. For a succinct analysis of these factors as they related to women in Mexico, see M. Srinivasan, "Impact of Selected Industrial Technologies on Women in Mexico," in Women and Technological Change in Developing Countries, ed. Rosalyn Dauber and Melinda Cain, pp. 89-108 (Washington, D.C.: American Association for the Advancement of Science, 1981).

13. Kuzmin defines a stagnant microsystem as being "characterized by a kind of 'low level equilibrium' whose distinguishing feature

is that usually no economic surplus worth mentioning is produced." See S. A. Kuzmin, "An Integrated Approach to Development and Employment," International Labour Review 115 (May-June 1977): 330.

14. Roger Revelle, "Energy Use in Rural India," Science 192 (June 4, 1976): 969.

## REFERENCES

Cecelski, Elizabeth. 1979. The Role of Electrification in Development. Washington, D.C.: Resources for the Future.

Copper, J. F. 1976. "Taiwan." In Energy Policies of the World, edited by Jerry Mangone, vol. 3. New York: El Sevier.

Dalal, Vikram L. 1977. "Energy Strategies for India: A Technological Perspective." Paper prepared at the Center for Environmental Studies, Princeton University, Princeton, N.J., July.

"Energy Technology Assessment: Consideration of Geographical Scale." 1977. Journal of Energy and Development, vol. 2, no. 2 (Spring).

India, Planning Commission. 1976. Fifth Five Year Plan, 1974-79. New Delhi: Controller of Publications.

Kuzmin, A. S. 1977. "An Integrated Approach to Development and Employment." International Labour Review, vol. 115, no. 3, (May-June).

Makhijani, Arjun. 1977. "Energy Policy in Rural India." Economic and Political Weekly (August).

National Academy of Sciences. 1976. "Energy for Rural Development." Report of an Ad Hoc Panel of the Advisory Committee of Technology Innovation Board on Science and Technology. Washington, D.C.: NAS.

National Commission on Science and Technology. 1976. Report of the Expert Panel on Solar Energy. Government of India.

Parikh, Jyoti. 1978. Assessment of Solar Applications for Transfer of Technology: A Case of Solar Pumps. Laxenburg, Austria: International Institute of Applied Systems Analysis, December.

Prasad, R. Krishna, and A. K. N. Reddy. 1976. "Technological Alternatives and the Indian Energy Crisis." National Seminar on Energy, Hyderabad, March.

Ramakumar, R. 1977. "Technical and Socio-Economic Aspects of Solar Energy and Rural Development in Developing Countries." Solar Energy, vol. 19.

Reddy, A. K. N. 1978. "Energy Options for the Third World." Bulletin of Atomic Scientists, May.

Revelle, Roger. 1976. "Energy Use in Rural India." Science, vol. 192 (June).

Revelle, Roger, et al. 1967. "Water and Land." In The World Food Problem, vol. 2, Washington, D.C.: U.S. Government Printing Office.

Sankar, T. L. 1976. "India." In Energy Policies of the World, edited by Jerry Mangone, vol. 3. New York: El Sevier.

Sen, Lalith K. 1979. "Rural Energy Scene in India." In Productivity, vol. 19, no. 4 (January-March).

Shah, Narottam. 1971. "Natural Resources in the Indian Economy." In The Natural Resources in the Indian Economy, edited by Vadilal Dagli. Bombay: Vora.

"The Social Costs of Energy Choices." Christianity and Crisis, vol. 38, no. 15 (October).

Solar Energy Research Institute. 1979. Annual Review of Solar Energy, vol. 2. Boulder, Colo.

Spears, J. S. 1978. "Wood as an Energy Source: The Situation in the Developing World." Paper presented at the 103rd Annual Meeting of the American Forestry Association, October 8, 1978.

Srinivasan, Mangalam. 1981. "Impact of Selected Industrial Technoloties on Women in Mexico." In Impact of Selected Industrial Technologies on Women in Developing Countries, edited by Rosalyn Dauber and Melinda Cain. Washington, D.C.: American Association for the Advancement of Science.

_____. 1976. "Integrating Women in Agricultural Development." Program and Training Journal of Peace Corps, vol. 4, no. 2 (September).

United Nations, General Assembly. 1981. "Report of the Synthesis Group." Preparatory Committee for the United Nations Conference on New and Renewable Sources of Energy, 3d sess., March 30-April 17.

Usmani, I. H. 1979. "Energy Banks for Small Villages." Bulletin of the Atomic Scientists 35 (June): 40-44.

Von Oppen, M. 1977. "The Sun Basket." Appropriate Technology 4 (November): 8-10.

# 5

# A FRAMEWORK FOR ASSESSING TECHNOLOGICAL DEVELOPMENT: THE CASE OF MEXICO

VIVIANE B. MÁRQUEZ

## NOTE ON TECHNOLOGICAL DEVELOPMENT

The concept of technological development is difficult to define when it refers to society as a whole as opposed to specific developments. Thus, technological development includes such different aspects as the level of scientific and technological knowledge and know-how as they are institutionalized by academic and research organizations. It also includes the level of industrial apprenticeship prevalent in industrial enterprises and the speed with which technological changes are adopted and diffused throughout the system, as well as the degree to which the state instrumentalizes such changes. Technological development should not be confused with innovativeness, as it is possible to find some technologically developed societies producing few innovations. Technological development, therefore, refers to the set of dimensions and processes that contribute to increasing the level of scientific knowledge and technical know-how in current use in a given society. Moreover, it characterizes a society as a whole as opposed to individual units within that society.

The discussion of the scientific and technological leadership of developed versus developing countries has slowly matured over the past two decades from very broad diagnoses to more complex and specific analyses supported by an increasingly richer and more accurate data base. Thus, general discussions of the "technological gap" have been succeeded by those of "technological dependency," in turn divided into narrower issues, such as the transfer of technology, appropriate technology, the building of a scientific and technological infrastructure, the role of transnational corporations, the role of the state, and the choice of technologies. This phenomenon has been paralleled by a gradual learning process during which initially unduly optimistic or simplistic diagnoses of the situation of developing countries have been discarded. At the same time, there has been a growing awareness that the problem goes far beyond the analytical confines of economics or engineering where it has been initiated and, therefore, has to be confronted with the tools of other disciplines from sociology to political science, from psychology to history.

Initially defined as a specific problem requiring the involvement of specialists, the persisting scientific and technological lag of developing countries has now come to be considered a global problem intricately interwoven with all facets of social and economic organization in these countries. As a consequence, what is now called for—indeed, what has been a major concern of scholars and policy makers during the past few years—is the establishment of a new methodological strategy allowing for such a global view to be effectively incorporated into research.

Technology assessment[1] may be regarded as the promising new analytical arsenal somewhat hurriedly assembled and consecrated in response to this demand for a holistic approach to science and technology. The promise it holds is nothing less than synthesizing the various fragments of information available into one coherent whole eventually to be handed out to planners and policy makers.

At this stage, it may be too early to evaluate this new effort. Nevertheless, it is not too early to warn about its probable pitfalls. In the past, the common lot of multidisciplinary approaches has been the superficial and eclectic accumulation of knowledge with little internal unity and little more to look back upon than the building up of erudition (Weisbecker 1978). Contrary to common belief, "bringing together" is not equivalent to "fitting together"; neither does one necessarily lead to the other. What is needed to fulfill the promises of multidisciplinarity in technology assessment, as elsewhere, is a methodology to articulate the different analytical perspectives brought together into a consistent whole.

This chapter shall attempt to establish some general methodological guidelines for technology assessment in order to reconstruct the totality of interconnected processes that intervene in (or interfere with) the process of technological development in Third World countries, selecting Mexico as a case for illustration. It shall propose an analytical scheme that represents a first step toward clarifying the nature of the relationships between different levels of analysis and disciplinary perspectives commonly adopted in the discussion of this problem.

The general argument that will be presented is that Mexico has acquired through its historical experience (shared in part with some other Latin American countries) certain structural and dynamic characteristics that act as constraints to technological development, where the latter is construed in the broadest possible fashion to include the concept of "technological capacity" proposed by the United Nations,[2] as well as the concept of "technological autonomy" adopted by Mexico's major scientific agency, the Consejo Nacional de Ciencia y Tecnología (CONACYT).[3]

The first part shall critically review common diagnoses that have been proposed on the subject of the generation of technological development in Latin America, pointing out the limitations of unidisciplinary perspectives and their tendency to consider all manifestations of a problem as relevant to their explanation and, hence, covered by proposed remedies. The chapter shall then proceed to propose a broad analytical scheme that allows for a better articulation between these separate diagnoses and shall finally analyze the Mexican situation as a case study that illustrates the points presented.

## OBSTACLES TO THE CREATION OF TECHNOLOGICAL DEVELOPMENT IN LATIN AMERICA: A BRIEF REVIEW

Almost every scholar who has recently taken stock of the state of the art in scientific and technological progress in Latin America (for example, see Sábato [1980]) will agree that we are a great deal wiser now on such matters than we were at the onset of the discussion in the early 1960s. We now know how ubiquitous the problems of science and technology are and no longer believe that they can be solved by merely spending more money on research or the training of more people abroad. Nevertheless, despite such sophistication gained in the conceptualization of science and technology issues in Latin America, it is important to face some of the basic shortcomings of past analyses. Should these not be recognized, the potential fruitfulness of a global approach might be seriously limited. Therefore, at the risk of being unfair to more than one, we shall venture some broad generalizations regarding current diagnoses of the obstacles to the creation of technological development.

One generalization that may be proposed is that discussions of scientific and technological deficiencies in Latin America have mostly consisted in identifying problems serially rather than interrelatedly. Thus, we are being constantly confronted with long lists of apparently discrete items, such as the following: domestic research in Latin American countries is imitative; domestic research in Latin America is not linked to productive apparatus; local scientific and technological activities suffer from low productivity; the demand of local productive systems is oriented toward foreign markets; technology transferred to Latin America has been laborsaving (Sábato 1980, Sagasti and Gerrero 1974; United Nations 1979). In short, there has been a wealth of analyses but little effort to integrate identified obstacles into a coherent frame of reference.

There are some clear advantages to having compiled an accurate list of all ills brought about by the misuse of science and technology common to Latin American countries throughout the past two decades. As such, it may be regarded as an achievement considering the broad and necessarily vague statements about dependency that were the starting point. Nevertheless, this empirical eclecticism does present the problem of ordering and recognizing under different manifestations the same basic features. Should this be achieved, the tasks of analysis and policy making would be greatly facilitated. It may be, for example, that one is basically talking about the same thing when we mention the low productivity of local research and development, the brain drain, and the difficulties of channeling foreign-trained professionals and scientists back into research activities. Again, one may be talking about the same thing when we simultaneously deplore the unwillingness of Latin American governments to take strong measures to promote science and technology (Herrera 1971, 1972; Wionzcek 1980) and comment upon the contradictions between public measures designed to encourage research and development and those designed to promote industrial activity (Unger and Márquez 1981).

The second point is that current discussions usually fail to define a hierarchical order among phenomena. Thus, for example, it may not be helpful to consider statements about the low quality of domestic research and development as being on the same analytical level as the orientation of local industry demand for research and development toward foreign sources. By doing so, one might erroneously be led to believe that it is the low productivity of local research and development that has "turned off," so to speak, local demand, when in fact, local research and development only emerged late imperfectly and ineffectively because the local industries' appetite for foreign products had already been satisfied by import substitution policies.[4]

Another way to establish a general hierarchy among observed phenomena might be to postulate the primacy of one given analytical level over another. One might, for example, posit the problem of political order[5] as theoretically determinant of the form and institutional strength of ongoing science and technology arrangements in a given society. In that way, resistance to change in these arrangements would be sought in consideration of vested interests, instead of the usually alleged inefficiencies and inadequate human resources.

A third problem is the failure to distinguish between levels of analysis, so that, for example, the actions of industrial firms are commonly regarded as mere "transmission belts" of the international technological market situation without granting these intervening units any possibility of modifying such broad schemes. The logic of this position leads to equally untenable extremes commonly found implicit in many discussions: either business leaders are defined as unaware of macroeconomic forces and, therefore, are mere puppets acting out these forces, or they are supposed to know what they are doing and, therefore, consciously conspire to perpetuate technological dependency by making all the socially undesirable decisions.

The problem for businessmen, however, is not to choose between moral responsibility and blissful ignorance, but to make decisions that will ensure the survival and success of their enterprise. Presently in most Latin American countries with a private or mixed enterprise system and tariff protection, technological wisdom varies drastically from the single businessman's perspective to that of a science and technology regulating agency. In most of these countries, the best technology for the entrepreneur from a cost, as well as an organizational, point of view is usually that which beats the union, beats inflation, and secures the firm's position in the market. That means that the same entrepreneur will find it more attractive to buy foreign technology that gives him exclusive representative rights on the local market, purchase the most powerful machinery available with the help of tax and import duty rebates (thereby creating idle capacity and increasing costs), and hire as few people as possible.

On the other hand, from a national standpoint the best technology will be one that creates employment, saves capital, fosters local innovation and adaptation, and uses optimally local resources. If the two perspectives are contradictory, it is a problem for policy making, not one of persuading the entrepreneur that he is really wrong and that science and technology regulating agencies or academic economists know what is best for him.[6]

One last shortcoming in current analyses of science and technology issues is the generalized tendency to favor the broad and macroanalytical over the individual and exceptional. This tendency is a natural consequence of the lack of firsthand empirical research in Latin America and the proliferation of secondary analyses usually based upon highly aggregated data.[7] As a result, policy measures that aim at the general are often defeated by the particular, and the potential for social change stemming from leadership and deviant cases is often missed. This tendency, however, has been recently reversed as a result of the work of various research projects supported by several international organizations.[8] In particular, some of these works have shown that many individual firms are capable of "beating the system" in the sense of pursuing adaptive changes and engaging in innovative activities despite the strong structural odds against them (James 1973; Márquez 1979; Pérez Acevez 1979). This would indicate the appropriateness of developing policies that are relevant to such individual units rather than always to legislate for the majority.

## ELEMENTS TOWARD A GENERAL ANALYTICAL FRAMEWORK FOR THE ASSESSMENT OF TECHNOLOGICAL DEVELOPMENT

The preceding discussion has shown that the basic problem faced by technology assessment is not so much to seek new facts as to link up those already accumulated without committing the fallacies of either reducing macroanalytical explanations to microanalytical explanations or reifying microphenomena into simple instances of broad forces.

In order to achieve this objective, one would have to start with the basic postulate that the obstacles to reaching technological development in Latin America do not act as sequentially ordered causal chains (where A leads to B, B leads to C, and C in turn leads to technological dependency) but have simultaneous effects at different levels. Furthermore, one would have to assume that these obstacles interact systemically, creating the familiar vicious circles whereby, for example, infrastructural science and technology deficiencies are reinforced by institutional weaknesses (science and technology policies). These in turn provoke individual deviations from the inadequately established rules, thereby further increasing the difficulties of building up a science and technology infrastructure likely to promote technological development.

One way of ordering these facts that might prove theoretically fruitful, as well as have positive practical consequences for policy making, might be to draw a distinction between policy-specific levels of analysis. In other words, one would separate analyses and arguments (in order later to join them) as a function of the kinds of policy measures that would be called for in order to bring some change into the situation, bearing in mind that the effectiveness at one level does not necessarily resolve problems at other levels (according to the initial postulate).

A simple taxonomy of policy-specific levels that might be a first approximation to the stated goal would consist of three levels of analysis. The first is the broadest macro level at which national science and technology policies interfere with other sectoral policies and/or international involvements, so that these would have to be modified in order to secure strong science and technology measures. This first level will be called political because of the strong tendency of its elements to act as zero-sum games wherein any single policy has a potential for bolstering the power of some interest groups and weakening others. This perspective is also political in the sense that changing the situation would involve a great deal of bargaining by the state with different power centers.

The next intermediate level of analysis would be that at which policies attempt to alter the behavior of industrial firms, laboratories, and research institutes by means of policies applying to specific groups of institutions. This level will be called institutional in the sense that measures proposed at this level would aim at fostering good scientific and technological practices (from a national standpoint) on the part of these units taken as wholes so as to achieve overall improvement (for example, measures aimed at controlling foreign investment, reducing royalties, and promoting research). This analytical level would be naturally complementary of the preceding level, insofar as altering the behavior of individual units would sometimes involve changing the value of environmental technological parameters for these units so as to make poor technological management costly. Such changes would necessarily involve adjustments with other sectoral policies.

The third-lowest level of analysis deals with individual actors engaged in science- and technology-related activities, such as research and development, local innovations and adaptations, or the diffusion of innovations (for example, extensionism in rural areas). The purpose of policies directed at this level would be to improve the conditions in which individuals can pursue these kinds of activities so as to prevent the chronic brain drain endemic to developing countries and transform exceptional and unplanned innovative activities into planned and institutionalized ones incorporated into the global strategy of development. This level of analysis will be referred to as the level of individual action owing to its emphasis upon individual and exceptional contributions to technological development.

## STRIVING FOR TECHNOLOGICAL DEVELOPMENT: THE CASE OF MEXICO

In Mexico, as in most countries of Latin America, technological development and its systematic linkage with the productive system have not occurred as spontaneous social processes. In spite of being one of the most industrialized countries of that region and despite much effort to induce changes of a scientific and technological nature by state intervention, Mexico remains a technologically dependent society.

Our purpose in reviewing the case of Mexico is not to make an evaluation of the science and technology policy instruments developed and implemented to date in that country, which has been done elsewhere (Nadal 1977) but to review the major science and technology problems identified in the literature. In doing so, the three levels of policy analysis defined in the preceding section of this chapter will be used with a view to identifying policies relevant to each level of analysis. Nevertheless, the discussion that follows should not be considered exhaustive but merely indicative of a methodology.

The singular importance of state intervention in the Mexican economy has generally been recognized (Bennett and Sharp 1979; Cordera 1979; Godau and Márquez in press; Pires 1979; Solis 1972; Tello 1979). Not only has the state designed major economic and social parameters within which private enterprise must function, but it has also acted directly in the economy through the numerous state and parastate institutions under its jurisdiction.[9]

State control over the private sector essentially manifests itself through credit mechanisms, the regulation of imports, and various public services. In the first case, the state may exercise total discretion over the granting or withdrawing of credit, as well as the flow of money, thereby determining priorities in productive investment on the basis of and as defined by national interests. Additionally, the state has the ability to encourage or restrain imports by alternatively lowering or raising tariff barriers. At the same time, the state should, in principle, be able to directly influence quality improvements in domestic production, as well as exert pressures in order to eliminate inefficiencies in the national productive apparatus. In the same manner, the state issues science and technology policy directives through CONACYT, which are presumably executed by the various educational and research institutions in the country.

Evidently, this de jure description of the state's theoretical capacity to intervene in the national economy on the basis of established policy instruments does not project a very realistic image. It is therefore important to discover the reasons behind the relative moderation on the part of the state in its role of intervention, precisely because this is what would make it possible to understand the somewhat modest place occupied by science and technology policies, as compared with other facets of state intervention in Mexico.

In order to identify the obstacles to a clearly defined or continuous action on the part of the Mexican state in the interest of science and technology, it is necessary to consider the whole spectrum of policy instruments at the disposal of its public agencies in relation to the social and economic constituencies to which they are directed. The main problem appears to be one of conflict between the different roles played by the state. Such conflicts arise out of the contradictory tensions exerted by the different interest groups affected by these policies, all of which are included in the postrevolutionary political coalition (Cardoso 1978; Collier 1979; Kaufman 1973; Malloy 1976;

Reyna 1976; Reyna and Weinert 1977). This situation is evidenced, for example, in the political struggles between private business and political spokesmen of the large impoverished urban and rural masses. Rather than being played out outside of the state (for example, via parties or legislative bodies), these struggles tend to be reflected in the inner workings of the state, among other things, in the form of parallel policy packages destined to placate the demands of each politically incorporated group taken separately. This, in turn, creates conflicts between different facets of state intervention and different branches of government, which have sometimes been erroneously blamed on inadequate bureaucratic organization.

For example, there are contradictions between the role of the Mexican state as the promoter of industrial development and its role as generator of employment. In the first case, the state accepts virtually any investment, no matter how technologically inadequate or how little adapted to the basic needs of the economy; in the second it stresses the necessity for the establishment of industries that save on capital and provide employment for the least skilled out of the most economically depressed regions.

At the same time, the creation of employment has paradoxically been the alleged reason for sustaining economically inefficient or technologically inappropriate enterprises. Therefore, any firm, regardless of its technological makeup, represents the generation, no matter how modest, of employment, whose disappearance could, in addition to aggravating unemployment, lead to unwanted pressures from labor leaders. As a consequence, its survival becomes a political imperative, far more so than its efficient or technologically appropriate management.

Likewise, there are contradictions between the state's endorsement of conventional import substitution and protectionist policies on the one hand and its support of science and technology policies aiming at the limitation and control of the flow of foreign technology and materials on the other. In its role as the creator of industrial development, the Mexican state has developed since World War II a series of instruments designed to create a favorable investment climate, which was especially needed after the nationalization of the oil industry and electric utility companies. These same economic stabilizing measures have at the same time created adverse conditions for the development of science and technology and the achievement of full employment.

During the presidency of Luis Echevarria, this type of contradiction was evident as equal pressures were exerted in favor of distributive and redistributive policies.[10] This meant that the government acted equally strongly in favor of industrial investment and of social measures designed to increase the standard of living of the poor masses. As a result, Mexico witnessed an unprecedented high level of inflation coupled with an acute confrontation of government with private business (Arriola 1976; Godau and Márquez in press) and various forms of social unrest (for example, land invasions by campesinos).

In this general framework, the science and technology policies that followed the creation of CONACYT must be understood primarily as ancillary to social and economic measures rather than policy instruments in their own right. For example, the package of regulatory measures that were enacted in order to control and reorient foreign investment contains some elements of science and technology policy. That was the case for the establishment of the National Registry of Technological Transfer or for the General Law on Trademarks and Patents. Nevertheless, all of these measures were more strongly oriented toward economic policy than toward science and technology policy. In effect, their main purpose was to lighten the burden of some payments in royalties and imported inputs quite independently of the establishment of general norms related to science and technology. On the other hand, it was not until the end of 1980 that some mechanisms were created in order to encourage industrial firms to pursue activities of research and development.[11] Apart from this modest attempt, still barely incipient in 1981, no systematic efforts have been made in order to orient the demand for technical and administrative assistance toward internal sources or to discourage industrial firms from overcapitalizing. In other words, state intervention during that period was limited to reducing the heaviest financial costs of technological dependency with little regard for establishing general conditions for encouraging technological development.

If we now examine the present sexennium (1976-82), its emphasis upon production and reconstruction of the economy becomes evident. In order to achieve this objective, a relaxation of tensions between government and the private sector was needed. This seems to be the meaning of President Lopez Portillo's Alliance for Production program, which at the same time as it wooed the private sector and attracted savings and investment back to Mexico provided an assurance of political stability to foreign creditors.

In such circumstances, it was unlikely that any science and technology policy liable to bring about conflict with private investors would be pursued very vigorously. In fact, it can be asserted that in spite of great international salience of scientific and technological issues, chiefly owing to the 1979 UN conference in Vienna, science and technology have taken second place in Mexican politics since 1977 (Wionczek 1980).

In sum, the general political climate in Mexico during the past ten years, but especially since 1977, has prevented the establishment of clearly defined or strongly enforced independent science and technology objectives. This, in turn, results from the fact that technological development objectives have not been supported by any political constituency capable of exerting effective pressures upon government and that the stability of the system has depended heavily on its ability to offer attractive conditions to private enterprise.[12] As a result, science and technology policies have been consistently constrained by other sectoral or international policies despite efforts to create comprehensive plans for the development of science and technology (CONACYT 1976, 1978).[13]

Therefore, one may conclude that the present science and technology situation in Mexico cannot be entirely explained by the conditions of supply of technology from the international market, but it also results from the action of the state. This action, evidenced in tariff, fiscal, financial, and employment policies, has transformed factor prices to an extent that is determinant of the preference of business firms for inappropriate technologies (from a national standpoint).

It also follows from this discussion that the reversal of this situation might involve a serious confrontation with those vested interests that are in flagrant contradiction with the stated goals of technological development and could, therefore, require exceptionally strong and well-coordinated mechanisms of implementation in order to be made robust enough to withstand such generally adverse conditions.

If we now turn to the second analytical level, namely, that which concerns the behavior of institutional actors (for example, industrial firms, state enterprises, research laboratories, and industrial consulting firms), the talk consists of evaluating the degree to which these contribute to promoting or preventing technological development and in what ways policy instruments may "reach" down to them. The important methodological point is not to focus upon institutional behavior merely as a dependent variable of the general processes of political economy that have just been outlined, but to conform to our original assumption that each policy-specific level of analysis has its own dynamic in addition to overall linkages with levels above and below.

In Latin America, a great deal of interest has focused upon studies of technological innovation in advanced economies and their diffusion to Third World countries via traditional mechanisms of technology transfer and transnational corporations. In particular, neoclassical economic theory has been criticized for its neglect of technology as an explanatory factor (Urquidi and Nadal 1979).

Out of such a critique has emerged a common concern for identifying technology as an endogenous variable related to its own environmental conditions, resulting in attempts to build microeconomic models of investments in research and development (Freeman 1974) and to define the characteristics of the markets of technological transactions between industrial firms of advanced countries and those of developing countries. This debate resulted in the redefinition of technology as another factor of production that may be appropriated and finally offered on the market (Arrow 1969; Cooper and Hoffman 1978; Tillet 1976; Vaitsos 1974) with special emphasis placed upon technology transfers incorporated in direct foreign investment in Third World countries (Fajnzylber and Martinez Tarrago 1976; Michalet 1976; Stanzik 1972; Sunkel 1973; Wionczek 1971, 1976; Wionczek, Bueno, and Navarrete 1974).

If one now considers this general situation from the perspective of the individual industrial firms, the alternatives it would face in principle would be either to develop its own technology or to buy it. Nevertheless, the most common option in Mexico, as in most developing countries, has been the purchase of foreign technology. Therefore, economic analysis in purchasing firms has been de facto reduced to a comparison between the costs of imported technology and the benefits the firm would derive from using them, a major one being the monopoly it acquires over the use of imported innovations.[14] The social cost excluded from such calculus is the consequent long-term inhibition of the potential for technological development in a developing country, resulting in the constant widening of the technological gap separating it from advanced economies, which in turn consolidates it into dependent status in the world economy.

The analysis of administrative decision making in industrial firms (merely assumed but not analyzed in economic theory) has been incorporated into the classical theory of organizations; as its name indicates, it has deep affinities with neoclassical and classical economic theories. Initially, its approach consisted of reducing the behavior of industrial firms to a series of decisions, where the latter were understood as conscious and rationalized procedures for maximizing results as a function of preestablished objectives, ultimately profit.

Later, this approach was to be modified by, among other things, demonstrating the impossibility of the clarity of objectives or the full rationalization of means to attain them (March and Simon 1958) or the oversimplicity of the profit maximization principle (Sfez 1975). Thus, it was shown that industrial enterprises often take nonprofitable decisions for strategic purposes (Chandler 1969; Thompson 1967) and have to face many different types of environmental pressures that cannot be reduced to simple cost and benefit calculations.[15]

Despite such developments, the influence of classical organizational schemes can nevertheless be felt in some of the explanations proposed for the behavior of industrial firms in Third World countries. Typically, when the apparent inefficiencies of the industrial process in these countries are not attributed to macroeconomic processes, they are reduced to problems of individual rationality or training,[16] thereby disregarding organizational processes that cannot be resolved either by general national policies or by individualized management training.

The alternative is not to make the individual firm or research institution in some sense responsible for the general situation but to identify problems that would be more easily analyzed and treated at the institutional level. A case in point is the problem of fostering research and development activities that create the basis for genuine technological development. In most developing countries that problem has usually been treated at the broadest macroanalytical level of educational policy or human resources (Anderson and Bowman 1965; Blaug 1972; Hanson and Braembeck 1962; Pinera and Selowsky 1976).

Mexico is no exception to that generalization. As in most dependent societies, science and technology are not vitally linked to the units of industrial production or economic decisions but form isolated pockets that are kept artificially alive, so to speak; few institutions in the surrounding society can derive any benefits from their activities.[17] In industrial firms, research and development is not considered an endogenous process related to the totality of the firm but a separate specialty to be carried out exclusively by highly specialized cadres in specialized institutions. As a result, the general absence of institutionalized research and development in Mexico is attributed to the lack of qualified personnel and/or lack of vision on the part of industrial entrepreneurs.

For example, a great deal of effort and resource has been devoted to training cadres, with little notion that once they are trained they will be absorbed by more prestigious institutions capable of offering them career development and leadership. Meanwhile, understaffed, underequipped, underorganized, underpaid, and overpoliticized research institutions proliferate without ever reaching their critical mass or amounting to any more than collections of small individual projects with little relevance to major decision centers (Carvajl and Lomnitz 1981).

Such efforts only show the persistence of dependence upon imported models of development with little relevance to local conditions, even in the very process of attempting to overcome these conditions. Rather than count upon generalized economic or human resources policies in order to stimulate indigenous research, more thought should be given to the institutional conditions in which research is carried out, and new organizational formulas should be proposed (Carvajal and Lomnitz 1981). For example, in the industrial sphere, rather than attempt to duplicate existing models of research and development from industrialized countries, the Mexican state might do better to create new institutional forms of technological development likely to connect better with the existing industrial structure, such as funding research and development institutions specialized in adapting foreign technology to local conditions or encouraging the pooling of funds on the part of groups of private firms in order to finance their own research and development.

In fact, Mexico has already experimented with such formulas on a small scale with very encouraging results. For example, the Instituto Mexicano de Investigaciones Tecnológicas has been operating with public funds for several decades, executing pilot projects for Mexican industry. Recently, this institution has expanded its activities, but its scale remains small. Other more recent institutions are the Instituto Mexicano del Petroleo or the Laboratorios Nacionales de Fomento Industrial. On the other hand, since 1970 a government-supported office under the jurisdiction of CONACYT has acted as a kind of industrial extension, representing an important source of inexpensive engineering advice for Mexican firms. Nevertheless, there is presently no comprehensive evaluation of such vitally important social experiments that would be indispensable for future policy. In addition, the scale on which they operate is too modest to have any determining weight upon the economy or to allow any predictions about the probable consequences of their generalization.

Where agricultural production is concerned, much effort has been spent in Mexico on demonstrating on a pilot basis the feasibility of a large range of technologically innovative procedures adapted to local conditions (Márquez 1981). In particular, new formulas have been sought in order to improve the economic position of rural communities by creating small-sized agro-industrial firms. These have been capable, with minimum capital and human resources, of processing intermediate products (such as unrefined sugar or packet meat) as inputs to larger capital-intensive firms, thereby allowing for the coexistence and linkage of different levels of technology for different needs within the same economy.

An additional aspect of technological development that has received scant attention until recently is the so-called process of industrial apprenticeship. This leads to the third policy-specific level of analysis, namely that of individual or small-group action within industrial firms, conducive to what has been called minor innovations and adaptations of imported technology (Katz 1976).

There are a number of case studies in Mexico demonstrating the existence in isolated individual firms of a great deal of individual innovative capacity, know-how, and ability to make optimal technological choices (Márquez 1979a), even in the absence of specific firm-based decisions or in the presence of adverse environmental conditions. One learns, for example, about the head of a Mexican firm who created a technology for producing a chemical product in high demand out of cheap local raw material, but whose efforts have been nevertheless countered by a large state firm using imported technology (Pérez Acevez 1979). Another Mexican firm created technology for producing paper out of sugarcane bagasse, but later ended up merging with a transnational corporation because it was unable to stand on its own (Márquez 1979b; Strassman 1968). Another case study reveals unusual skill deployed by the head of a medium-sized paper mill in northern Mexico in selecting and modifying used machinery based upon the establishment of a worldwide network of information (James 1979). Finally, a number of firms in the food product industry have learned to produce parts and modify components, not as a result of systematic firm policies but as spontaneous outcomes of the activities of their repair shops (Unger and Márquez 1981).

Although much interest has been spurred recently about these kinds of studies under the leadership of several international institutions (for example, the University of Sussex' Science Policy Research Unit, the International Development Research Center of Canada, the World Bank-Economic Commission for Latin America projects, and UNESCO), much still remains to be learned from the unused potential of technological development in developing countries that does not correspond to traditional research and development functions. In particular, little is known about the conditions in which creative individuals can be retained in research activities, what kind of leadership is needed for research institutions, or in what ways these can be made to connect meaningfully with industrial firms.

## SUMMARY AND CONCLUSIONS

Among the tasks confronting technological assessment in the developing countries, this chapter discussed the impact that common technology transfer practices from industrialized to Third World countries have had upon the latter's potential for technological development, taking the case of Mexico as a central point for illustration.

One major difficulty in technology assessment, if it is to be truly multidisciplinary, is obtaining a coherent picture out of divergent perspectives on the same problem. A threefold division among political, institutional, and individual processes, corresponding to situations calling for different analytical and policy instruments, was proposed.

This division was based upon the assumption that relatively autonomous social processes would take place at each level, so that applying correctives at any one level would offer no guarantee of radical change at the next. The intervening phenomenon that explains how these different levels are linked up, yet harbor relatively autonomous dynamics, is power. By that, we mean that technology is not simply a solution for problems of production that is more or less profitable or more or less adjusted to a given situation. It is fundamentally a resource around which different interconnected power centers are created and will act so as to maintain their position should any threatening change occur.

We have argued, for example, that the import-substitution path to economic development chosen by so many developing nations, and by Mexico in particular, gives strong political leverage to sectors whose maintenance in power depends upon perpetuating technological dependence. At the same time, such policies are unlikely to allow for the rise of strong political constituencies in favor of technological development capable of exerting pressures upon their governments in order to reverse the political options associated with import substitution.

The first conclusion that may be drawn at this level of analysis is that it would be unwise to expect major changes to take place in the near future. Nevertheless, this does not mean that governments cannot begin to select areas marked for future technological development and change general political conditions for these areas only. In the case of Mexico, for example, the very heterogeneity and unconnectedness of the action of the state would give it sufficient flexibility so that pinpoint action would not introduce major disequilibrium into the system.

General political conditions will determine processes at the next level down, insofar as they are conducive to the building up of an industrial structure dominated by large firms (usually transnational) whose continued leadership in the market depends upon the constant incorporation of imported innovations. These leading firms, in turn, constrain the behavior of other firms (in the same industrial categories), which find themselves unable to compete successfully unless they adopt similar strategies.

A second conclusion would be that given the buildup of these trends over the years, they are unlikely to be reversed by merely offering piecemeal incentives to technological development. In effect, such measures would be tantamount to only slightly modifying general political conditions without altering the industrial power structure that they have created. Likewise, encouraging research and development in specialized institutions is unlikely to lead to fundamental changes in the market strategy of leading firms or to convince other firms to compete with new untested weapons. In both cases, the risks involved would by far outweigh potential gains.

Last, if one considers the conditions in which individuals acting within this structural framework will have to operate, we should conclude that it is not enough for them to be appropriately trained in order to become successfully incorporated into institutions likely to blaze new trails in technological development. Although much has been made about the inadequacy of foreign training for personnel from developing countries (accusing returnees from these programs of adding cognitive dependence to the structural dependence of their home countries), more should be investigated about the inadequacy of research institutions in these countries. They are typically apathetic, unproductive, and riddled with internal politicking and represent far greater obstacles to technological development. In view of the political and institutional setups discussed earlier, it becomes evident that the problems of these returnees would be far greater if they came back intent upon upsetting the technological status quo rather than merely aspiring to the usual prestigious technocratic position. One would, therefore, conclude that unless special measures are taken in order to make scientific and/or technological development worth their while, the exceptionally trained or gifted will continue to be diverted away from science and technology (excepting bureaucratic posts in that area) and will be absorbed by sectors more likely to provide them with satisfying working conditions and chances for social promotion.

As a global conclusion that may be derived from this analysis, one might add that the complexity of phenomena that confronts technology assessment is such that it will not prove sufficient in the future to merely appraise static situations in vitro, so to speak. What is called for are pilot experiments that combine different policy-specific levels in order to test interdependences and find the right combinations to reverse established processes. Also needed, prior to such conditions, is the will to change and the preparedness to engage in a sustained effort, which may ultimately be the most crucial ingredient of technological development.

## NOTES

1. It is not our intention to deal with the problem of the definition of technological assessment, however needed that may be. Consequently, we shall use the definition proposed by the Seminar of Technology Assessment held in Bangalore, India, from October 30 to November 10, 1978. "Technology assessment is the process for systematic analysis, forecasting and evaluation of the broad range of impacts on society pertaining to technological change and choice in order to identify public policy options" (Standke and Korn 1979).

2. The UN General Assembly, at its Seventh Special Session in 1975 in calling for the World Conference on Science and Technology of Vienna, defined its objectives in the following way:

> A United Nations Conference on Science and Technology for development should be held in 1978 or 1979 with the main objectives of strengthening the technological capacity of developing countries to enable them to apply science and technology to their own development; adopting effective means for the utilization of scientific and technological potentials in the solution of development problems of regional and global significance, especially for the benefit of the developing countries; and providing instruments of cooperation to developing countries in their utilization of science and technology for solving socio-economic problems that cannot be solved by individual action, in accordance with national priorities. [Standke and Korn 1979, emphasis added]

3. In the third chapter of the first National Plan of Science and Technology issued by CONACYT, technological autonomy is defined as follows:

> La autodeterminación tecnológica no supone la autarquía en éste camp, sino la libertad de decisión en la búsqueda, selección, negociación, utilización, asimilación, adaptación, y generación de tecnología; esto es, el desarrollo de una capacidad para establecer relaciones tecnólogicas sin dependencia. [Consejo Nacional de Ciencia y Tecnología, 1976, p. 41]

4. This does not mean, naturally, that said low productivity may not feed back upon an already weak demand, thereby perpetuating the situation.

5. Political order being broadly defined as the familiar "who gets what, where, and when."

6. Such wrongheaded persuasion attempts are basic to recent efforts on the part of various international associations to create what has been called "technological management" for the benefit of businessmen. This new, would-be branch of business administration literally aims at teaching business leaders how to make appropriate technological decisions from a national standpoint, implicitly disregarding the contradiction between macroperspective and microperspective in that respect.

7. It is also a consequence of the great vogue for survey studies in the 1960s and early 1970s, which greatly devalued case studies considered "unrepresentative."

8. See, for example, the research supported by the International Development Research Center, such as Sagasti 1978; Nadal 1977; Mer-

cado 1980; Boon 1978; Rattner 1978, 1981; and Unger and Márquez forthcoming. See also the research supported by the IBRD, ECLA, IDRC, and UNDP: Katz 1976, 1978a, 1978b; Katz and Cibotti 1978; Maxwell 1978; Pérez Acevez 1979; Vitelli 1978; and Sercovich 1979.

9. The term paraestatal in Mexico refers to institutions that belong to the public sector but are nevertheless financed and managed autonomously. The expression public sector, in turn, refers to governmental agencies, institutions operating under public grants (fideicomisos), and public enterprises. In contrast, the term private sector refers to any privately owned concern.

10. The tendency of Mexican politics to alternate between emphasis upon production and that on redistribution from one sexennium to the next has been coined the "pendulum theory" (Needler 1971, pp. 46-49). The peculiarity of the Echeverria administration was in the fact that both polarities were emphasized simultaneously.

11. A Presidential Decree published in the Diario oficial of November 26, 1980, grants some fiscal advantages to institutions that, owing to their research and development activities, become members of the National Registry of Scientific and Technological Institutions under the jurisdiction of CONACYT. This decree was enforced starting in 1981, which means that most institutions that would qualify are now barely applying to become members of that registry. Before 1981, no tax write-offs were granted on the basis of research and development, resulting in the fact that most industrial firms did not even separate such activities on their books (Unger and Márquez forthcoming).

12. This is especially true since the mid-1970s, during which time inflation has worsened Mexico's position on the international market. The guarantee of tariff protection, fiscal advantages, and inefficient pursuit of antipollution policies, therefore, become added attractions with which to woo private investors.

13. This situation is not limited to science and technology policies. Examples of contradictory policies are abundant in Mexican politics for reasons briefly proposed. For example, the systematic protection of industry has meant that large agro-industrial complexes have been allowed to pursue strategies that are contrary to the interests of rural populations equally included in agricultural policies. Another example is the recognition of the need to foster preventive medicine, which has been countered by the strongly institutionalized support of the pharmaceutical industry, thereby reinforcing traditional curative modes or organizing public health.

14. This term is taken in the sense given to it in Schumpeter (1928).

15. This approach is known under the name of contingency approach in organization theory. For more details, see Hall 1972, chap. 10.

16. We are referring here to the arguments of modernization theory that assign causal effects to personality traits of individuals categorized as either modern or traditional.

17. This does not mean that such institutions are necessarily parasitic or that their findings could not be used if general conditions were different. What it means is that their activities are not a vital part of the dynamic of the surrounding society, which depends upon foreign sources of technological development incorporated into imported technology.

## REFERENCES

Anderson, C., and M. Bowman, eds. 1965. Education and Economic Development. New York: Aldine.

Arriola, Carlos. 1976. "Los grupos empresariales frente al estado." Foro internacional 16 (April-June): 449-95.

Arrow, J. K. 1969. "Classificatory Notes on the Production and Transmission of Technological Knowledge." American Economic Review: Papers and Proceedings.

Boon, J. K. 1978. "International Technology Transfer Dynamics: The Market of Polyester, Textile and Apparel Technology." Noordwijk aan Zee, Holland. Mimeographed.

Bennett, Douglas, and Kenneth Sharp. 1979. "El estado banquero y empresario: el caracter de última instancia de la intervención economica del estado Mexicano, 1917-1970." Foro internacional 20 (July-September): 29-72.

Blaug, M., ed. 1972. Economía de la educación. Madrid: Editorial Tecnos.

Bueno, G. In press. "El desarrollo tecnológico: sus relaciones con la evolución de América Latina." Comercio exterior, vol. 31, no. 5.

Cardoso, F. H. 1978. On the Characteristics of Authoritarian Regimes in Latin America. Cambridge: Center of Latin American Studies, At the University Press.

Carvajal, Raul, and L. Lomnitz. 1981. "El desarrollo científico en México: Es posible multiplicarlo con los mismos recursos?" Ciencia y desarrollo, March-April, pp. 90-98.

Chandler, Alfred D. 1969. Strategy and Structure. Cambridge, Mass.: MIT Press.

Collier, D., ed. 1979. The New Authoritarianism in Latin America. Princeton, N.J.: Princeton University Press.

Consejo Nacional de Ciencia y Tecnología. 1978. Programa nacional de ciencia y tecnología, 1978-1982. Mexico: CONACYT.

_____. 1976. Plan nacional indicativo en ciencia y tecnología. Mexico: CONACYT.

Cooper, C. Y., and K. Hoffman. 1978. "Transactions in Technology and Implications for Developing Countries." SPRU and University of Sussex, England. Mimeographed.

Cordera, Rolando. 1980. "Estado y desarrollo en los '80's: modelo para armar." In Economía y Política en el México actual, edited by José Maria Calderon, Sergio de la Peña, S. Cordera, R. Olmedo, and M. Labra. Mexico: Ed. Terranova.

_____. 1979. "Estado y economía en México: La perspectiva histórica." Economía de América Latina, September, pp. 101-23.

Fajnzylber, F. Y., and T. Martinez Tarrago. 1976. Las empresas transnacionales: expansion a nivel mundial y proyeccion en la industria mexicana. Mexico: Fondo de Cultura Económica.

Freeman, C. 1974. The Economics of Industrial Innovation. Harmondsworth, England: Penguin.

Godau, R., and Viviane B. Márquez. In press. "Estado, sociedad y la empresa privada en México." Revista de administracao de empresas.

Hall, Richard. 1972. Organizations: Structure and Process. Englewood Cliffs, N.J.: Prentice-Hall.

Hanson, J., and C. Braembeck, eds. 1962. Education and the Development of Nations. New York: Holt, Rinehart and Winston.

Herrera, Amilcar. 1972. "Social Determinants of Science Policy in Latin America: Explicit Science Policy and Implicit Science Policy." Journal of Development Studies, vol. 9, no. 1 (October).

_____. 1971. Ciencia y política en América Latina. Mexico: Siglo XXI.

James, D. 1973. "Used Automated Plants in Less Developed Countries: A Case Study of a Mexican Firm." Interamerican Affairs 27 (Summer): 31-46.

Katz, Jorge. 1978a. Cambio tecnológico, desarrollo económico y las telaciones intra y extra regionales de la América Latina. IBRD/ECLA Program of Research on Science and Technology, Monograph no. 30 BID/CEPAL/BA, Buenos Aires.

_____. 1978b. "Creación de tecnología en el sector manufacturero argentino." Trimestre económico 45 (January-March): 167-90.

_____. 1976. Importación de tecnología, aprendizaje e industrialización dependiente. Mexico: Fondo de Cultura Económica.

Katz, Jorge, and Y. R. Cibotti. 1978. "Marco de referencia para un programa de investigación en temas de ciencia y tecnología en América Latina." Trimestre económico 45 (January-March): 139-65.

Kaufman, Susan. 1973. "Decision Making in an Authoritarian Regime: Theoretical Implications from a Mexican Case Study." World Politics 26 (October): 28-84.

Kennedy, C. 1974. "Induced Bias in Innovation and the Theory of Distribution." Economic Journal 83 (September): 98-157.

Malloy, James, comp. 1976. Corporatism and Authoritarianism in Latin America. Pittsburgh: Pittsburgh University Press.

March, James G., and H. Simon. 1958. Organizations. New York: John Wiley & Sons.

Márquez, Viviane B. 1979a. Dinámica de la empresa mexicana: perspectivas políticas económicas y sociales. Mexico: El Colegio de México.

_____. 1979b. "Organizational Structure and Innovativeness in the Pulp and Paper Industry of Mexico." In Technological Progress in Latin America: The Prospects for Overcoming Dependency, edited by James H. Street and Dilmus D. James, pp. 149-72. Boulder, Colo.: Westview Press.

Márquez, Viviane B., ed. 1980. "Situación perspectiva de la tecnología adecuada para el desarrollo apropecuario en México." Paper prepared for the Experts Meeting on the Relation between Science, Technology and Employment in Rural Areas, Bogotá, December 8-12, UNESCO Document SC/80/Conf. 662/3.

Maxwell, Philip. 1978. "Estrategia tecnológica óptima en un contexto económico difícil: la evolución de la planta siderúrgica de ASINBAR en Rosario, Argentina." Trimestre económico 45 (September-November): 1033-64.

Mercado, A. 1980. Estructura y dinamismo del mercade de tecnología industrial en México. Mexico: El Colegio de México.

Michalet, Charles-Albert. 1976. "The International Transfer of Technology and the Multi-National Enterprise." Development and Change 7: 157-74.

Nadal, Alejandro. 1977. Instrumentos de política científica y tecnológica en México. Mexico: El Colegio de México.

Needler, Martin. 1971. Politics and Society in Mexico. Albuquerque: University of New Mexico Press.

Perez-Acevez, Luis Alberto. 1979. "Innovación, productividad y comportamiento tecnológico de una empresa mexicana." In Dinámica de la empresa mexicana: perspectivas políticas económicas y sociales, edited by Vivian B. de Márquez. Mexico: El Colegio de México.

Piñera, Sebastian, and Marcelo Selowsky. 1976. "The Economic Cost of 'Internal' Brain-Drain: Its Magnitude in Developing Countries." International Bank for Reconstruction and Development, Bank Staff Working Paper no. 243.

Pires, P. 1979. "Estado y economía: dos perspectivas teóricas." Foro internacional 20 (July-September): 1-28.

Rattner, Henrique. 1977. "Gestão tecnológica." Revista de administracao de empresas, vol. 17 (November-December), special issue on technological management.

Rattner, Henrique, ed. Forthcoming. Gestão tecnológica na industria de alimentos latino-americana: estudo comparativo. São Paulo: Fundacao Getuilo Vargas.

Reyna, J. L. 1976. "Estado y authoritarismo." Neuva politica, vol. 1 (April-June).

Reyna, J. L., and R. S. Weinert, comps. 1977. Authoritarianism in Mexico. Philadelphia: Institute for the Study of Human Issues.

Sabato, J. 1980. "Desarrollo tecnológico en América Latina y el Caribe." Revista de la CEPAL, no. 10 (April), pp. 87-100.

Sagasti, Francisco R., and M. Guerrero C. 1974. Desarrollo científico y tecnológico de América Latina. Buenos Aires: BID/INTAL.

Salter, W. E. G. 1960. Productivity and Technical Change. Cambridge: At the University Press.

Samuelson, P. A. 1965. "A Theory of Induced Innovation along Kennedy-Weisacker Lines." Review of Economics and Statistics, November.

Sercovich, F. 1979. Design Engineering and Endogenous Change: A Summary. Buenos Aires: BID/CEPAL.

_____. 1975. Tecnología y control extranjero en la industria Argentina. Buenos Aires: Siglo XXI.

Sfez, Lucien. 1974. Critique de la décision. Paris: A. Colin.

Solis M., Leopoldo. 1972. "El Sistema financiero en 1980." In El perfil de México en 1980, edited by David Ibarra, Ifigenia M. de Navarrete, Leopoldo Solis, and Victor L. Urquidi, vol. 1, pp. 73-88. Mexico: Siglo XXI.

Standke, K. H., and A. Korn. 1979. "Technology Assessment in Planning for Development." In Science and Technology in Development Planning, edited by V. L. Urquidi, pp. 99-106. New York: Pergamon Press.

Stanzik, Karl Heinz, and H. H. Godoy, eds. 1972. Inversiones extranjeras y transferencia de tecnología en América Latina. Santiago de Chile: ILDIS/FLACSO.

Sunkel, Osvaldo. 1973. "The Pattern of Latin American Dependency." In Latin America in the International Economy, edited by Victor L. Urquidi and Rosemary Thorp. London: Mcmillan.

Sylos-Labini, P. 1969. Oligopoly and Technical Progress. Cambridge, Mass.: Harvard University Press.

Tello, Carlos. 1979. La política económica en México, 1970-1976. Mexico: Siglo XXI.

Thompson, James. 1967. Organization in Action: A Behavioral Science Approach. New York: McGraw-Hill.

Tillett, A. 1976. "Propiedad y patentes: el caso de México." Comercio exterior 26: 908-39.

Unger, Kurt, and Viviane B. Márquez. Forthcoming. La tecnología en la industria alimentaria Mexicana: diagnóstico y procesos de incorporación. Mexico: El Colegio de México.

United Nations, General Assembly. 1976. Preparatory Committee for the United Nations Conference on Science and Technology for Development. 3d sess., January 22-February 2.

Urquidi, Victor, and A. Nadal. 1979. "Algunas observaciones acerca de la teoría económica y el cambio técnico." Trimestre económico, vol. 46, no. 182 (April-June).

Vaitsos, C. 1974. Intercountry Income Distribution and Transnational Enterprises. Oxford: Clarendon Press.

Vitelli, Guillermo. 1978. "Cambio tecnológico, estructura de mercado y ocupación en la industría de la contrucción Argentina." Trimestre económico 451 (September-November): 997-1032.

Weisbecker, L. W. 1978. Technology Assessment: Broadening Development Decisions. Paper delivered at UN Seminar on Technology Assessment for Development in Bangalore, October 30-November 10.

Wionczek, M. 1980. "Las desventuras de la política de ciencia y tecnología en México." Foro internacional, no. 421 (March).

_____. 1976. "Notes on Technology Transfer through Multinational Enterprises in Latin America." Development and Change 7: 135-55.

_____. 1971. Inversión y tecnología extranjera en América Latina. Mexico: Joaquin Mortiz.

Wionzcek, Miguel, G. Bueno, and J. E. Navarrete. 1974. La transferencia internacional de tecnología: el caso de México. Mexico: Fondo de Cultura Económica.

# 6

# HYDROPOWER DEVELOPMENT IN THE PEOPLE'S REPUBLIC OF CHINA: ECONOMIC AND SOCIAL IMPACTS

CHENG XUEMIN

## HISTORICAL AND GEOGRAPHICAL BACKGROUNDS

### General Features

With more than 4,000 years of uninterrupted written history, the People's Republic of China (PRC) has one of the oldest civilizations in the world. Some of the early achievements in Chinese literature and science are well known to all. For many centuries, China's civilization, and its economy, led the world. However, the industrial revolution that stimulated a rapid growth of modern economies in Western countries in the eighteenth and nineteenth centuries failed to take place in China. As a result, China gradually became the target of military invasions and economic exploitation by aggressive capitalist countries. From the middle of the nineteenth century to the middle of the twentieth century, the modern history of China had been marked with humiliation and suffering, as well as uprising and revolution.

By land area, the PRC, with 9.6 million square kilometers, is the third largest country in the world. It spans 5,200 kilometers from east to west and 5,500 kilometers from north to south. It is a land with a widely diversified climate and an extremely complex topography. Generally speaking, the western and northwestern parts of the PRC are cold and dry and are largely occupied by deserts, mountain ranges, plateaus, and loess regions. The eastern and southeastern parts of the PRC are mild and wet and are blessed with large fertile plains and hilly areas. The total cultivated land in the PRC, mostly in the east and southeast, is only $15 \times 10^8$ mu (100 million hectares), which accounts for only 7 percent of the world's total of cultivated land area. More than 90 percent of the total population, and consequently, the economic activity, is concentrated in the eastern and southern parts of the PRC.

Agriculture has long been the principal economy in the PRC. More than 80 percent of its population are peasants. Prior to the founding of the People's Republic of China in 1949, widespread famines caused by natural hazards, as well as by years of unending warfare, were frequent to Chinese people. Modern industry was insignificant at that time. Most of the war-supporting heavy industries in the northeastern provinces built by the Japanese during their occupation there from 1931 to 1945 were subsequently dismantled by Soviet troops. The light industries in the coastal cities, such as Shanghai, Tianjin, and Guangzhou, were dwindling as a result of the war. In 1949 the newly founded People's Republic of China found itself economically less developed than the excolonial nation of India and the recently defeated Japan.

## Economic Developments in the Past Three Decades

The new government lost no time in rebuilding the war-torn economy in the PRC. Things appeared to be going smoothly until the successful completion of the five-year plan in 1957. A number of mistakes were made afterward and caused a lot of criticism. In spite of this, statistics still show remarkable progress in economic development during the past three decades. Table 6.1 indicates the changes of some of the basic indexes over a number of specific years.

The population of the People's Republic of China has doubled during the past 30 years. There was serious debate on the population problems in the PRC in the 1950s. It is regrettable that the debate was not treated fairly, and the right opinion was not accepted by the authorities until very recently. The chaotic conditions that prevailed during the Cultural Revolution made things even worse and resulted in an increase of nearly 200 million people between 1965 and 1975. Very tight control measures have been adopted by the government since 1978. Such measures were pushed forward with difficulty but were considered to be absolutely indispensable. In many instances, the PRC stands as a giant as far as the total sums are concerned, but when divided by the denominator of population, it becomes a dwarf.

The increase of the population by 1.8 times from 1949 to 1980 has been fortunately compensated for by an increase in food production of 2.8 times during the same period. Admitting that the food consumption of 324 kilograms per capita per year is a very low standard, the PRC cannot afford to live on imported food, nor can the world food market afford to feed a large population like the PRC's. The only way out is to rely on its own people and to work hard on its own land. This has proved to be successful.

TABLE 6.1

Statistics of Basic Indexes

| | 1949 | 1952 | 1957 | 1965 | 1975 | 1980 |
|---|---|---|---|---|---|---|
| Population (in millions) | 541.67 | 574.82 | 646.53 | 725.38 | 919.70 | 982.55 |
| Food production (in millions of tons) | 113.20 | 163.90 | 195.05 | 194.55 | 284.50 | 318.22 |
| National income (in billion yuan of renminbi) | 35.8 | 58.9 | 90.8 | 138.7 | 250.5 | 363.0 |
| Products (in billion yuan of renminbi) | 46.6 | 82.7 | 124.1 | 198.4 | 450.4 | 661.9 |
| Agricultural | 32.6 | 48.4 | 53.7 | 59.0 | 128.5 | 162.7 |
| Industrial | 14.0 | 34.3 | 70.4 | 139.4 | 321.9 | 499.2 |

Note: In 1980, 1 yuan of renminbi = U.S. $0.67.

Source: Almanac of China's Economy, 1981.

The patterns of the PRC's economy have changed drastically during the past three decades. In 1949 the ratio between the total value of agricultural products and that of industrial products was about 1:0.4; in 1980 this ratio has been changed to 1:3.07. Industry has become the leading sector in the PRC's economy. Before 1949 commodities like steel nails had to be imported from abroad. Nowadays, airplanes and electronic computers are manufactured in Chinese factories. Most of the industrial plants in the PRC are equipped with homemade machinery. Of course, not all of these products have achieved the highest technological level in the world. There is still a long way to go in the PRC's course of industrialization.

But the rapid growth in industry has brought about the deficiencies that are now being reviewed and that are the center of attention in the current readjustment and reform programs. Deficiencies within the industrial sector appear as the imbalance between heavy industries, which are engaged in producing productive goods (such as steel and machinery), and light industries, which are engaged in producing consumer goods. An overemphasis of the former and an ignorance of the latter have held up the pace of further economic development. Deficiencies also exist in the imbalance between the rapid expansion of industries and the long neglect of the improvement of people's living conditions. As a step toward a remedy, in recent years the government has begun to allocate a large portion of national resources to large-scale housing programs; wage reforms; and the expansion of education, health, and municipal facilities at the expense of slowing down industrial development.

The gross national income in the PRC is not calculated in the same way as the Western practice but is assessed at Y 369 RMB (U.S.$230) per capita in 1980.

## ENERGY RESOURCES AND THE ROLE OF HYDRO POTENTIAL IN THE PRC

### Principal Energy Resources

The People's Republic of China is abundant in energy resources, especially in coal and hydropower. The oil reserve in the PRC is also very promising, but the exact statistics are not yet finalized. The government policy is to limit the use of oil for power generation as much as possible. For reasons of simplicity, the following discussions will concentrate on coal and hydro.

The total geological coal reserve (known resources) in the PRC is estimated at 1,460 billion metric tons, and the measured coal reserve is assessed at 600 billion metric tons. Of this, about 380 billion tons are classified as thermal coal suitable for electric power generation. Assuming a recovery factor of 50 percent, it leaves a total economically recoverable reserve of thermal coal at 190 billion tons, or 136 billion tons of coal equivalent (heat value, 7,000 kilocalories per kilogram).

The total theoretical hydro potential of the PRC, estimated in a survey carried out from 1977 to 1980, is 676 million kilowatts, corresponding to 5,922 billion kilowatt-hours per annum. The total exploitable hydropower potential of 1,705 identified hydro sites is assessed at 378 million kilowatts, corresponding to 1,923 kilowatt-hours per annum.

The current fuel consumption of all thermal power plants in the PRC averages 413 grams of coal equivalent for each kilowatt-hour generated. Using a conversion factor of 400 grams of coal equivalent per kilowatt-hour, the total exploitable hydro potential in the PRC equals 770 million tons of coal equivalent per annum. Taking into consideration the renewable capacity of hydropower, a time period of 100 years is multiplied to make it comparable with the coal reserve. Thus the total exploitable hydro resources in the PRC equals 77 billion tons of coal equivalent.[1]

The overall situation of the PRC's energy resources can be summarized as follows:

1. The predominant energy source in the PRC at present, and for many years to come, is coal, which accounts for about two-thirds of the energy resources.

2. Hydro potential accounts for about one-third of the PRC's energy resources. As a clean and cheap source of energy, it is expected to play a more important role in the future.

3. Oil is not seen as having economic potential for power generation in the PRC in the foreseeable future. The oil-burning power plants now in operation will be converted to or replaced by coal-burning power plants in due course.

## Distribution of Coal Reserves

Coal reserves in the PRC are not evenly distributed. Nearly 70 percent of the total reserve is concentrated in northern China, especially in the Province of Shanxi and the Autonomous Region of Inner Mongolia. The two economic regions with a relatively higher level of industrialization, namely the eastern region and the northeastern region, possess less than 5 percent of the national coal reserve. All the other provinces in the central southern, northwestern, and southwestern regions have meager coal reserves with but a few exceptions. The transportation of coal is, therefore, an ever increasing problem in the PRC. The general direction of the flow of coal is from north to south. For example, the Province of Shanxi, which possessed one-fifth of the national coal reserve and produced one-sixth of the national coal production in 1980, has its six railway lines fully loaded to carry the coal out of the province. Any further expansion in coal production in this province is severely hampered by insufficient transport facilities. In 1978 about 40 percent of the PRC's railway transportation was for coal. The railway is by far the most important means of transportation in the PRC. However, coastal transportation is vital to the coastal cities. In 1979 more than 80 percent of the waterborne transportation for Shanghai was composed of coal and oil. The congestion in seaports makes the situation even worse.

Distribution of Hydro Potential

Since 1949 nationwide surveys on hydro potential have been conducted three times. The latest survey was carried out from 1977 to 1980, when the national interest turned to economic development. The recent survey has the advantage of the availability of a wealth of information accumulated from numerous reconnaissances, explorations, and planning studies of most of the important rivers in the PRC during the past three decades. In addition to the evaluation of the theoretical potential, a greater effort is being made to assess the exploitable potential for the recent survey.

It has been generally accepted in the PRC that a nationwide survey of exploitable resources can serve as a very useful guide to the setting up of regional planning and the formation of long-term development plans.

The results of the survey indicate that the hydro potential in the PRC is also unevenly distributed. Two-thirds of the total hydro potential is concentrated in the southwest. The potential in the northern, northeastern, and eastern regions accounts for less than 8 percent of the national total, but the absolute values still represent a significant amount.

The hydro sites surveyed are classified into categories according to the investigation and engineering work done to each site. There are a number of sites where considerable amounts of geological exploration and engineering design have been carried out, comparative to the requirements of feasibility studies. Others may need more investigation. This is very useful to the planning engineers and executives who want to know how many hydro sites are readily available for early implementation and how much investigation work is required for the further expansion of long-term plans. Arrangements may have to be made for such time-consuming exploration years beforehand.

The distribution of hydropower potential in the PRC is shown in Tables 6.2 and 6.3. The hydro potential of Taiwan Province is not included in the current survey.

The Yangtze is the largest river system in the PRC. It drains one-fifth of the national land area and contains more than one-half of the national hydro potential, mostly concentrated in the upper reaches. The Yellow River, although ranked second largest in drainage area, possesses only 6.1 percent of the total power potential. The annual runoff of the Yellow River is only one-twentieth of the Yangtze. The Yalutsangpo River and the southwestern rivers, which include the Lancan River (Upper Mekong) and the Nu River, possess a great amount of hydropower as they flow along or across the edge of the Xizang Plateau with a big discharge. There exist a number of excel-

TABLE 6.2

Hydro Potential in the PRC according to Region

| Region | Theoretical Potential (in millions of kilowatts) | Exploitable Potential: Millions of Kilowatts | Exploitable Potential: Billions of Kilowatt-hours | Percent of Total |
|---|---|---|---|---|
| Northern | 12.30 | 6.92 | 23.225 | 1.2 |
| Northeastern | 12.12 | 11.99 | 38.391 | 2.0 |
| Eastern | 30.05 | 17.90 | 68.794 | 3.6 |
| Central southern | 64.08 | 67.43 | 297.365 | 15.5 |
| Southwestern | 473.31 | 232.34 | 1,305.036 | 67.8 |
| Northwestern | 84.18 | 41.94 | 190.493 | 9.9 |
| Total | 676.05 | 378.53 | 1,923.304 | 100.0 |

Source: People's Republic of China, Ministry of Electric Power.

TABLE 6.3

Hydro Potential in the PRC according to River System

| River | Theoretical Potential (in millions of kilowatts) | Exploitable Potential: Millions of Kilowatts | Exploitable Potential: Billions of Kilowatt-hours | Percent of Total |
|---|---|---|---|---|
| Yangtze | 268.02 | 197.24 | 1,027.498 | 53.4 |
| Yellow | 40.55 | 28.00 | 116.991 | 6.1 |
| Pearl | 33.48 | 24.85 | 112.478 | 5.8 |
| Hai | 2.94 | 2.13 | 5.168 | 0.3 |
| Huei | 1.44 | 0.66 | 1.894 | 0.1 |
| Northeastern | 15.31 | 13.70 | 43.942 | 2.3 |
| Southeastern | 20.67 | 13.90 | 54.741 | 2.9 |
| Southwestern | 96.90 | 37.68 | 209.868 | 10.9 |
| Yalutsangpo | 159.74 | 50.38 | 296.858 | 15.4 |
| Inland | 36.99 | 9.97 | 53.866 | 2.8 |
| Total | 676.05 | 378.53 | 1,923.304 | 100.0 |

Source: People's Republic of China, Ministry of Electric Power.

lent power sites of immense magnitude, but their development would have to be postponed owing to their remoteness.

The uneven distribution of coal and the uneven distribution of hydro compensate for each other to some extent. There still are regions lacking both coal and hydro, such as the northeast and the east. It has always been a great challenge to the energy planners to draft a well-balanced energy development program among all the economic regions in the PRC.

## HYDROPOWER IN CONNECTION WITH NATIONAL AND REGIONAL ECONOMIC DEVELOPMENTS

### Need for Harnessing the Rivers

Most of the large rivers in the PRC have been notorious for their hazardous natures. Since the founding of the People's Republic, one of the national objectives has been the harnessing of the unruly rivers. Comprehensive planning for a whole river basin, contemplating multiple-purpose developments for flood control, irrigation, power, and navigation, has been formulated for many large rivers in the PRC. A number of key projects have been constructed and have rendered satisfactory service. Many of the river schemes are, however, still in the course of implementation.

The development of the Yellow River plan is a notable example. The Yellow River has a serious problem in controlling its severe floods, which caused dike breaks in two of every three years before 1949 and claimed heavy losses in life and property. On the other hand, it is also the main source of water supply for irrigation in the northwest and north and is needed to relieve the PRC's principal wheat-producing lands from frequent droughts. In 1955 comprehensive planning for the development of the Yellow River Basin was formulated. The basin planning includes 46 projects with a total capacity of 22 million kilowatts. A number of storage reservoirs were also planned. Among them, the Sanmenxia Project and the Liujiaxia Project are the most important. The Sanmenxia Project in Henan Province was built from 1957 to 1960 with a main purpose of flood control. Excessive silting occurred during the initial filling of the reservoir leading to large-scale reconstruction work, which took place from 1964 to 1970. The silting of the reservoir has since been practically checked, and the reconstruction proved to be a success. The project is now operating at a lowered reservoir level with a reduced generating capacity of 250 megawatts and has retained its capability to hold catastrophic floods, reduce the damage of ice jams downstream, and supply irrigation water to the farm lands in Henan and Shangdong provinces during the spring season. The Yellow River has been saved from dike breaks for 32 consecutive years since 1949, and the six hydropower stations in operation on the main stream of the river have an aggregate generation capacity of 2,338 megawatts.

A multipurpose basin planning for the Yangtze River has also been contemplated. The key project of the Yangtze River development is the Three Gorges Project on the main stem of the Yangtze, and in the vicinity of the city of Yichang, Province of Hubei. It is a huge project with 73 billion cubic meters of reservoir capacity; of this, 37 billion cubic meters will be allocated for flood control. This project will have a generating capacity of 25 million kilowatts. The very scale of the project, especially the large amount of capital outlay required and the large number of people to be relocated (well over a million), prevented it from early implementation. For many years, speculations have arisen as to whether this project should be constructed according to its proposed magnitude or some alternative should be adopted.

Many of the tributaries of the Yangtze are mighty rivers by themselves. Basin plannings for these tributaries have been compiled as well. By the end of 1980, seven major projects were constructed on the tributaries of the Yangtze with a total generating capacity of 3,268 megawatts, and with relevant benefits in flood control, irrigation, and so on.

The Huei River system in the eastern part of the PRC and the Hai River system in the north are generally situated on large, flat plains. The development plans of these two rivers have been focused on flood control, drainage, and irrigation. Dikes, flood channels, regulating barrages, and upstream reservoirs were constructed in large numbers in the basins of these two rivers. Hydropower development has been limited to medium-sized installations in the hilly upstream tributaries.

## Growth of Electricity Demands

The growth of electricity demands has been very rapid in the PRC since 1949. Total electricity generated in 1949 was 4.3 billion kilowatt-hours. It grew to 300.6 billion kilowatt-hours in 1980 and gives an average annual rate of growth of 14.7 percent. Hydropower has matched the general pace of development. It grew from 0.7 billion kilowatt-hours in 1949 to 58.2 billion kilowatt-hours in 1980, showing an average annual rate of growth of 15.3 percent.

Table 6.4 shows the growth of electricity and the percentage of hydro generation in the PRC in specific years.

TABLE 6.4

Growth of Electricity Supply in the PRC

| Year | Total Electric Generation (in billions of kilowatt-hours) | Hydro Generation (in billions of kilowatt-hours) | Percent of Total |
|---|---|---|---|
| 1949 | 4.30 | 0.70 | 16.3 |
| 1952 | 7.26 | 1.26 | 17.4 |
| 1957 | 19.34 | 4.82 | 24.9 |
| 1966 | 67.60 | 10.41 | 15.4 |
| 1970 | 115.86 | 20.46 | 17.7 |
| 1975 | 195.84 | 47.63 | 24.3 |
| 1980 | 300.63 | 58.21 | 19.4 |

Source: People's Republic of China, Ministry of Electric Power.

The percentage of hydropower compared with the total electricity produced has been varying between 15 percent and 25 percent. The highest hydro percentage was attained in 1957, when the First Five-Year Plan was successfully fulfilled. The hydro percentage fell afterward for various reasons. The total utilized hydropower in 1980 amounts to only 3 percent of the total exploitable hydro resources. This point has induced frequent discussions and criticisms within the related governmental and professional organizations in the PRC.

The breakdown of energy sources for electricity generation in 1980 is as follows:

| | Percent |
|---|---|
| Coal-burning plants | 59.4 |
| Oil-burning plants | 21.2 |
| Hydropower plants | 19.4 |

Coal is by far the most predominant source of fuel for electricity generation and is likely to maintain its role until the end of this century. The unusually high percentage of oil fuel for power plants is a result of the abnormal effects of the Cultural Revolution rather than a well-planned outcome. Nearly one-sixth of the annual oil

TABLE 6.5

Electricity Consumption for Different Uses
(in percent)

| Use | 1949 | 1980 |
|---|---|---|
| Agricultural | 0.6 | 14.9 |
| Industrial | 69.1 | 77.9 |
| Communication and transportation | 0.5 | 0.6 |
| Municipal and domestic | 29.7 | 6.6 |

Source: People's Republic of China, Ministry of Electric Power.

production in the PRC was burnt in power plants. It is considered to be a heavy loss in the nation's precious foreign exchange and a heavy burden to the national energy balance. Efforts are being made to improve the situation, but it needs both time and money.

The rate of growth of electric power supply outmatched the rate of growth of industrial production and that of the national income in the past three decades. However, electric power development seems still insufficient to meet the rapid growth in electric demands. As a result, power shortages have been reported in many parts of the country, and the shedding of electric supply has been unavoidable in certain cities, as well as in certain sectors of industry, especially the power-intensive industries.

Electricity consumption for different uses is shown in Table 6.5. The figures indicate that great changes occurred in economic structures in the PRC in the past 30 years. The rapid growth of electricity used in the agricultural sector is most remarkable. More than 80 percent of the electricity consumed in the rural areas in 1980 was for pumping irrigation, processing of agricultural products, and supplying factories run by the people's communes. Such consumers were simply nonexistent in 1949.

Industry holds the lion's share of electricity consumption over the past three decades with even more accentuation. Nevertheless, the internal structures within the industrial sector underwent substantial changes as well. For example, while the textile industry was the number one power consumer in 1949, this position is now held by the chemical industry. A large amount of electric power is used in the manufacturing of chemical fertilizers. The electricity consumption of the oil industry and the ferrous and nonferrous metallurgical industries also grew more rapidly than other industries. The declining trend of electric consumption in municipal and domestic uses indicates the unbalanced development in this area when compared with other economic sectors. Some of these trends indicate an advance in national economic development, while others are considered undesirable and are likely to have improved during the recent readjustment program.

The per capita consumption of electric energy in 1980 is only about 250 kilowatt-hours per annum, which is very low by global standards.

## Regional Developments

The geographical distribution of electricity generation among different economic regions in the PRC for 1952 and 1980 are shown in Tables 6.6 and 6.7, respectively.

In comparing Tables 6.6 and 6.7, one readily notes that the geographical distribution of electricity generation is better balanced in 1980 than in 1952. In 1952 the three coastal regions, namely, the north, northeast, and east, took about 89.3 percent of the nation's total electricity supply. By 1980 the share of electricity taken by the

TABLE 6.6

Distribution of Electricity Generation in 1952

| | Total Electricity | | Hydro | |
|---|---|---|---|---|
| Region | Millions of Kilowatt-hours | Percent | Millions of Kilowatt-hours | Percent |
| Northern | 1,269 | 17.5 | 0 | 0.0 |
| Northeastern | 2,958 | 40.7 | 1,193 | 40.3 |
| Eastern | 2,255 | 31.1 | 2 | 0.1 |
| Central southern | 436 | 6.0 | 0 | 0.0 |
| Southwestern | 274 | 3.8 | 63 | 23.0 |
| Northwestern | 69 | 0.9 | 2 | 2.9 |
| Total | 7,261 | 100.0 | 1,260 | 17.4 |

Source: People's Republic of China, Ministry of Electric Power.

TABLE 6.7

Distribution of Electricity Generation in 1980

| Region | Total Electricity Millions of Kilowatt-hours | Total Electricity Percent | Hydro Millions of Kilowatt-hours | Hydro Percent |
|---|---|---|---|---|
| Northern | 53,556 | 17.8 | 1,313 | 2.5 |
| Northeastern | 53,995 | 18.0 | 5,072 | 9.4 |
| Eastern | 83,816 | 27.9 | 9,018 | 10.8 |
| Central southern | 57,592 | 19.1 | 20,245 | 35.2 |
| Southwestern | 26,679 | 8.9 | 12,003 | 45.0 |
| Northwestern | 24,989 | 8.3 | 10,552 | 42.3 |
| Total | 300,627 | 100.0 | 58,211 | 19.4 |

Source: People's Republic of China, Ministry of Electric Power.

three coastal regions was reduced to 63.7 percent, while the share taken by the three inland regions rose from 10.7 percent in 1952 to 36.3 percent in 1980. The development of hydropower played an important role in this transition. As an example, in the central southern region hydropower generation rose from nil to 20,245 million kilowatt-hours in the 29 years. Hydro energy in the three inland regions occupied 35 percent to 45 percent of their total power supply in 1980. Admitting that electricity supply is one of the dominant factors in the general economic development, it can be concluded that the development of hydropower has made handsome contributions to a more evenly distributed national economy.

The development of hydropower has benefited the coastal regions, too. During the early years of the newly found republic, most of the hydropower stations built were in the coastal regions, where a ready power market was existing and expanding. The scale of development was mostly medium-sized, but it did stimulate economic growth in the regions concerned.

From the viewpoint of resources utilization, Table 6.8 shows the degree of utilization of hydro potential in different economic regions. It can be seen from Table 6.8 that in the coastal regions, a higher degree of utilization of hydropower had been attained, although with a smaller hydro potential. On the other hand, only a lower degree of utilization was achieved in the inland regions, despite greater hydro potential.

TABLE 6.8

Utilized Hydro Potential, 1980

| Region | Exploitable Potential (in millions of kilowatt-hours) | Utilized Potential (in millions of kilowatt-hours) | Degree of Utilization (in percent) |
|---|---|---|---|
| Northern | 23,225 | 1,313 | 5.7 |
| Northeastern | 38,391 | 5,072 | 13.2 |
| Eastern | 68,794 | 9,018 | 13.1 |
| Central southern | 297,365 | 20,245 | 6.8 |
| Southwestern | 1,305,036 | 12,003 | 0.9 |
| Northwestern | 190,493 | 10,560 | 5.5 |
| Total | 1,923,304 | 58,211 | 3.0 |

Source: People's Republic of China, Ministry of Electric Power.

Rural Electrification and Small Hydropower Stations

There are 800 million people in the rural areas of the PRC. They produce one-fifth of the world's foodstuffs with little modern machinery and maintain a very low standard of living. Sustained efforts have been made to improve the cultural and economic life in the rural areas of the PRC since 1949. Besides the land reform in the early 1950s and the organization of people's communes in the late 1950s, both of which have far-reaching effects, the supply of electricity is important progress in the development of the rural community. Prior to 1949, the percentage of the rural population in the PRC with an electricity supply was negligible. By 1980 about 50 percent of the rural population was enjoying the benefits of electrification. They are served by national power grids or supplied by small hydropower stations of their collective possession.

Lighting usually comes first when electricity is first introduced to a rural community. It brings about a big change in the daily lives of the peasants. Electric lighting extends daily activity for several hours that would otherwise be wasted. The availability of electricity opens the way to the introduction of cinemas, radio sets, and more recently, television sets, which enhance the rural cultural life magnificently. The processing of agricultural products comes next to lighting. Electric mills save a lot of time and labor at little cost and earn the appreciation of every rural household.

Irrigation pumping and drainage expanded drastically in the past decades. A number of large pumping stations were built in parts of the country. Some of the pumping stations consume as much as 100 megawatts of electric power. The major portions of small pumps are driven by diesel engines, but there is a tendency toward shifting from diesel engines to electric motors owing to restrictions in the use of fuel oil. Small industries are booming in the electrified rural areas. The most popular one is the small chemical fertilizer plant. Agricultural machinery factories come next. In addition, there is a large number of small industries in the rural areas producing coal, textiles, tobacco, iron, steel, and many items for daily use, according to the resources available locally.

Total electricity consumption in rural areas in 1980 was 37.4 billion kilowatt-hours, which accounted for 14.9 percent of national electricity consumption. Electricity supplied to the rural areas is charged at reduced prices.

There are 90,000 small hydropower stations in operation in the PRC at present with a total capacity of 6.68 million kilowatts. Among them are 1,700 stations with individual capacities of 500 to 12,000 kilowatts, having an aggregate capacity of 3.23 million kilowatts. The aggregate capacity of all the smaller hydropower stations less than 500 kilowatts each is 3.45 million kilowatts.

The construction of small hydropower stations is generally sponsored by local authorities with financial subsidy from the central government. The government subsidy was usually extended as a grant, but in recent years it has been more frequently financed by a loan, which will have to be repaid and bears a certain amount of interest. Maximum use is always made of local materials and the local labor force to minimize the construction cost. Manufacturing of small hydro turbines and generators has been very popular in the PRC. especially in the southern provinces. Small hydro units can be obtained conveniently and at a reasonable cost.

Many of the small hydro plants in the PRC are incorporated with irrigation facilities. The sharing of construction costs is definitely advantageous. However, conflicts in water uses sometimes arise. The issue becomes more acute when no storage reservoir is provided to compromise the conflicting water requirements. A storage reservoir is usually an expensive undertaking. Not all the small hydropower stations are fortunate enough to have one. Many of the small hydro stations have to be operated on a reduced supply or sometimes on an intermittent supply of electricity during the dry season. There has been a tendency for owners of small hydro plants to apply for integration into a nearby power grid. The problem is then determined by how much of the electric energy flow dependability can be absorbed by the power grid and how much the grid is willing to pay. In some southern provices, where the grid itself is composed of several unregulated hydro stations, the answer is not always favorable.

In view of the important role the small hydropower stations have played and are expected to continue to play in the rural electrification of the PRC, improvement in the performance of existing small hydropower stations is encouraged as well as the construction of new ones, aiming at a still higher degree of rural electrification and a more reliable and economical power supply to the vast Chinese rural community.

## TECHNOLOGY ASSESSMENTS

### River Basin Planning

Most of the important rivers in the PRC have been investigated on a river basin scale. In the very beginning, reconnaissance was usually made on the whole river course, identifying all potential sites for possible development. Geological explorations on key projects were carried out to various depths as the planning studies proceeded. Hydrological data were compiled not only from the records of numerous hydrometric stations equipped with modern instruments but also from historical documents or from water marks on important buildings or natural rocks, usually painted or even carved with the exact date of occurrence. Such water marks indicate the highest and occasionally the lowest water stage of rivers over the past several hundred years. Topographical maps are now obtainable from the national surveying and mapping organization, supplemented by detailed surveys of dam sites and reservoir areas by the design institute whenever necessary.

As previously stated in this article, multipurpose planning is generally favored for the development of Chinese rivers. The needs and problems of all relevant sectors of the national economy should be carefully considered. All the related organizations and departments should be duly consulted and invited to take part in the planning studies. A multiple-purpose development usually results in greater overall benefits, since all the functions concerned have a tendency to be complementary to each other. On the other hand, a multipurpose project usually requires a larger capital outlay than a single-purpose one. It is a setback in project financing, especially in a capital-insufficient country like the PRC.

A number of alternative overall development schemes for the whole river system are contemplated and optimized to give a complete picture of the ultimate development of the river basin. Such a scheme envisages the maximum extent of exploitation of the power resources, as well as other multiple-purpose benefits. The relative merits of each individual project can be evaluated in reviewing the overall scheme. A number of key projects or projects suitable for early development can also be sorted out. One of the principal targets of river basin planning is the choice of the first project to be constructed on a river system. This is usually a very important issue and would invite a lot of discussions and even debates. Anyhow, this is the first concrete result of a comprehensive river basic planning.

River basin planning thus formulated would be subject to reviews and revisions after a number of years of implementation. When the first project is nearing completion, another project would then be selected for construction according to the knowledge and experience accumulated and changing situations that may emerge with the lapse of time.

## Building of Reservoirs

In view of the erratic nature of Chinese rivers, it is necessary to build reservoirs to even out the high torrents in the flood season and to compensate for the insufficient supply of water during the dry season. A number of large reservoirs were built during the past 30 years.

The Xin'anjiang Reservoir in Zhejiang Province, with a total storage capacity of 22 billion cubic meters, is mainly a hydropower project with an installed capacity of 662.5 megawatts and relevant flood control and irrigation benefits. The reservoir caused inundation of about 20,000 hectares of cultivated land, and 270,000 inhabitants had to be relocated. The reservoir was constructed from 1957 to 1960.

The Danjiangkou Reservoir in Hubei Province, with a total storage capacity of 20.8 billion cubic meters, is principally a flood control project with a power installation of 900 megawatts. The reservoir, built from 1958 to 1964, caused inundation of 26,000 hectares of cultivated land, and 330,000 inhabitants had to be relocated.

The Sanmenxia Reservoir on the Yellow River in Henan Province, with a total storage capacity of 16.2 billion cubic meters, is a key project built from 1952 to 1960 for the harnessing of the river. Severe sedimentation occurred after its first filling and led to a large-scale reconstruction of its spillway system. The renovated project retained its functions in flood control, power generation, and irrigation on a reduced scale. The reservoir is now kept at low water levels in ordinary times, but will be used in case of an extraordinary flood. In building this reservoir, no less than 64,000 hectares of cultivated land were inundated and 320,000 inhabitants relocated.

The Xinfengjiang Reservoir in Guangdong Province, with a total storage capacity of 13.9 billion cubic meters, is a power project with an installed capacity of 292.5 megawatts and incidental flood control and irrigation benefits. The reservoir was built from 1958 to 1960, inundating about 12,000 hectares of cultivated land and relocating about 100,000 inhabitants.

The Liujiaxia Reservoir on the upstream of the Yellow River in Gansu Province, with a total storage capacity of 6 billion cubic meters, is the second key project (next to the Ganmenxia Reservoir) for harnessing the Yellow River as contemplated in the basin planning formulated in 1955. With an installed capacity of 1,160 megawatts, it is the largest hydropower station in operation in the PRC and, in addition, has important flood control and irrigation benefits. The reservoir was built from 1958 to 1969, including several interruptions, with about 5,000 hectares of cultivated land being inundated and about 30,000 people relocated.

The above mentioned are some of the outstanding examples. They give some idea of the scale of work confronted by Chinese hydro engineers. Summarizing the previous experiences, the following points might be cited:

1. Generally speaking, it seems unavoidable to build reservoirs on most of the Chinese rivers, which are too erratic to be efficiently utilized without a certain degree of regulation.

2. The size of the reservoir (the extent of flow regulation) should be carefully analyzed and well balanced to reach an appropriate choice. A reservoir of appropriate size is the one with an adequate storage capacity and causing least adverse effects in social and environmental aspects.

3. There is a tendency to build fewer and smaller reservoirs in the densely populated regions of the PRC. As a matter of fact, a number of reservoirs are built in reduced sizes compared with their original designs. However, in remote areas, such as Qinghai Province in the northwest, the Longyangxia Reservoir, with a total storage capacity of 20 billion cubic meters, is being built on the upper Yellow River, where relocation problems will be very minute.

4. Efforts have been made to exploit the most efficient use of existing reservoirs by developing successive projects in the downstream.

5. The reservoir relocation cost is escalating very rapidly, especially in recent years, owing to a general rise in the living conditions of the rural population in the PRC.

6. The work on reservoir relocation, although difficult and time consuming, can be done successfully with good organization and a high sense of responsibility (as proved by the relocation work achieved in the early 1950s).

## Dam Construction

There are 35 major dams with heights over 80 meters completed or presently under construction. They are classified into different types and are listed in Table 6.9.[2]

Concrete dams of the gravity type have been most popular in the PRC because of their simple design and construction. The Liujiaxia Dam on the upper reaches of the Yellow River in Gansu Province with a height of 147.5 meters tops the list of existing concrete gravity dams.

The highest gravity arch dam under construction is the Longyangxia Dam, also on the upper reaches of the Yellow River, with a height of 172 meters and an installed capacity of 1,280 megawatts. The highest double curvature arch dam under construction is the Dongjiang Dam in Hunan Province, with a height of 157 meters and an installed capacity of 500 kilowatts. The buttress and multiple arch dams in the PRC are considered as light-type dams; they are vulnerable to earthquakes. In several instances, severe cracks have occurred, and remedial measures have been necessary.

Although embankment dams are very common in the PRC in small projects, they occupy only a minor place in the high dam category. The difficulty of flood flow handling in spillway layout has prevented the adoption of this type of dam at many localities. Concrete diaphragm cutoff walls of 0.8 to 1.0 meters in width are frequently used in embankment dams on deep alluvial foundations. The maximum depth of concrete cutoff walls reaches 65 meters.

TABLE 6.9

Dams over 80 Meters in Height

| Type | Completed | Under Construction | Total |
|---|---|---|---|
| Generate gravity | 11 | 6 | 17 |
| Gravity arch | 2 | 3 | 5 |
| Double curvature arch | 2 | 2 | 4 |
| Buttress (diamond head) | 3(2) | 1 | 4 |
| Multiple arch | 1 | - | 1 |
| Embankment | 3 | 1 | 4 |
| Total | 22 | 13 | 35 |

Masonry dams are built in traditional ways in the PRC up to a height of over 100 meters. Most of them are on small rivers, and their power values are generally insignificant. A number of dams have been built in karstic limestone areas in the PRC. One example is the 164-meter-high Wujiangdu Dam (630 megawatts) in Guizhou Province. Karstic caves of various sizes were found up to a depth of 300 meters under the dam foundation, which had to be treated by an extensive and complicated grouting curtain system.

A special design was used for the Maotiaohe No. 4 Dam in Guizhou Province. The river bed of the dam site is covered with boulder and gravel deposits 27 meters deep. An arch bridge was designed to span over the river channel, and a 55-meter-high double curvature arch dam was built on the bridge. Two concrete diaphragm cutoff walls, each 0.8 meters thick, were built through the gravel overburden into the rock foundation to stop the seepage.

Spillway arrangements are one of the most important issues of dam design in the PRC. A number of concrete gravity/gravity arch dams have been designed with powerhouses inside the dam in order to leave sufficient space in the river channel for an overflow spillway utilizing the full length of the dam crest. The Fengtan Dam in Hunan Province, with a height of 112.5 meters and an installed capacity of 400 megawatts, is a typical example. Another example, though of a somewhat different design, is the Xin'anjiang Dam; 105 meters high, it is 662.6 megawatts in capacity. In this case, the powerhouse is located immediately below the top of the dam, and the spillway, with a flood discharge capacity of 13,200 cubic meters per second, is located right over the top of the powerhouse. This project has been in operation since 1960.

## Hydropower Machineries

Nearly all types of hydraulic turbines have been used in the PRC, including Pelton wheels, Francis turbines, Kaplan turbines, and bulb units. All the hydraulic turbine designs are standardized and are unified into a series of models covering a wide range of heads and capacities. Sometimes a number of similar hydro plants adopt the same turbine model to facilitate manufacturing. About 85 percent of the turbine units in operation are supplied by domestic manufacturers, while the remaining 15 percent are imported.

The majority of hydropower plants in the PRC are of medium heads, between 35 meters and 100 meters; thus Francis turbines are the most common type used. The largest Francis turbine unit in operation is the 300-megawatt unit installed in the Liujiaxia power station, with a rated head of 100 meters and a runner diameter of 5.5 meters.

It was manufactured by the Harbin Electric Machinery Works in Heilongjiang Province. Francis turbines, rated at 72.5 megawatts at a head of 73 meters, have been used in Xin'anjiang, Zhexi, and Xinfengjiang hydropower stations with but minor modifications. The Longyangxia station on the Yellow River will have four Francis turbines 320 megawatts each at a head of 120 meters. The Lubuge station, under construction in Yunnan Province, is designed to use four Francis turbines of 150 megawatts each at a head of 315 meters.

High head projects at the No. 3 and No. 4 stations of the Ilihe cascade scheme in Yunnan Province utilized a head of 629 meters each and installed eight Pelton wheels of the same design of 36 megawatts each, representing the highest head ever developed in the PRC.

Low head turbines, such as the 36-megawatt Kaplan units rated at 18 meters of head, have been used in the Qingdongxia, Bapanxia, and Wailongshan Stations. Kaplan units of 50 megawatts rated at a head of 14 meters were used in Fuchunjiang and Xijin hydro stations. The big Gezhouba hydro station is installed with 170-megawatt and 125-megawatt Kaplan units with a rated head of 18.6 meters. Three imported bulb units of 18 megawatts each at a head of 6 meters are under erection in Mazitan hydro station in Hunan Province.

Serious abrasion and cavitation of the hydraulic turbines owing to heavy silt content of the river flow are particular problems in the PRC's hydropower development. The runners have to be coated with antiabrasive materials for protection. A number of investigations on this issue have been conducted in recent years, and satisfactory results have been obtained.

In 1980, an experimental tidal power station was put into operation in Chekiang Province. A reversible bulb turbine, rated at 500 kilowatts at a head of 4 meters, is supplied by a domestic manufacturer and is operating now.

## NOTES

1. Adapted from Zhu Chengzhang, "The Role of Hydro Potential in China's Energy Resources," 1981, unpublished paper in Chinese.

2. Adapted from Li Eding, "Dam Construction in China," unpublished paper.

# 7

# ASSESSMENT OF TECHNOLOGY AND FIXED ASSETS OF INDUSTRY IN THE ECONOMIC DEVELOPMENT OF YUGOSLAVIA

MOMIR CECEZ

Science and technology, through their dynamism, have become active participants and contributors to economic development and the growth of nations. New methods and techniques, new devices and machines, and new aids to production have created new capital assets in all economic sections. Although the effects of science and technology are widely represented in industry, other sectors of modern life have also been revolutionized. The application of technology in industry has come to be synonymous with creating wealth and achieving economic development in most societies of the contemporary world.

By saying this, we should not neglect the mutual linkages and dependencies that exist between the development of an industry and the development of other activities, especially transportation, agriculture, trade, and the like.

Economic theories and models of development have played successful roles in the development of individual countries. A variety of conditions and the availability of development factors within these individual countries have confirmed the fact that there is no one model of development for all countries. At the same time, the empirical evidence indicates that the experience of one country can be advantageously studied by other countries. For this reason, this presentation of the data on Yugoslavia, together with a brief analysis of the contribution of fixed assets and technological progress in Yugoslav industry, may be useful from the standpoint of both economic theory and practice. An additional reason for such a belief is the fact that Yugoslavia has succeeded in transforming itself from an undeveloped agricultural country to a country with a gross national product (GNP) of nearly $3,000 per capita during the period of only the last 30 years.

Yugoslavia was an undeveloped country whose dominant activity was agriculture (72 percent of its population earned their income from agriculture in 1940). The agricultural economy of Yugoslavia was generally characterized by small land holdings, poor agricultural techniques, and very low returns. As in many other underdeveloped countries, the level of unemployment was very high, while labor was very cheap. Added to these factors was the devastation brought on by World War II, fascism, and the subsequent national liberation struggle. The enthusiasm of the Yugoslav people, based on a unity and decisiveness in wanting to build the country, played a most important role in the restoration of the destroyed economy.

The orientation toward a planned market economy and the development of a socialistic self-managing system were of special importance for the successful economic development and industrialization of Yugoslavia. By determining clear goals and choosing appropriate means for their realization, Yugoslavia has been achieving a dynamic growth rate during the last three decades. It has been making important structural changes as a consequence of predetermined economic development goals.

Yugoslavia is a country with diverse natural resources. Although the country does not have an abundance of any particular resource, it has adequate amounts of hydroenergy, coal, mineral ores, agricultural land, forests, and wood resources, as well as suitable conditions for the development of tourism.

The industrialization policy of Yugoslavia is based on existing industrial capacities, which have served both as the starting basis for industrialization and as a source of accumulation. The agricultural sector has been another source of accumulation for industrialization. Accumulation from agriculture was moving into industry through price policies, and this has had a negative impact on the development of agriculture during the first decade. Foreign resources and technology in the form of imported equipment, licenses, industrial cooperation, and joint ventures have been additional sources of capital accumulation and economic development.

## ECONOMIC DEVELOPMENT OF YUGOSLAVIA

Dynamic growth marks the main feature of Yugoslav economic development during the last 30 years (1950-80). The annual rise of the social product[1] in these years has averaged 6.3 percent. Significant structural changes have also occurred. The average annual rise of industrial production has been 8 percent during the same period, causing industrial production in 1980 to be 12 times greater than in 1950. The development of agriculture, however, has been at

TABLE 7.1

Trends in the Social Product, Based on 1972 Prices
(in percent)

| | Growth Rates | | | |
|---|---|---|---|---|
| | 1951-80 | 1951-60 | 1961-70 | 1971-80 |
| Social product total | 6.3 | 6.5 | 6.3 | 5.8 |
| Social sector | 7.4 | 6.8 | 7.5 | 6.4 |
| Individual sector | 3.4 | 5.6 | 2.2 | 2.4 |
| Industry | 8.8 | 11.2 | 8.1 | 7.4 |
| Agriculture, fishing, and water management | 3.4 | 5.9 | 1.6 | 2.6 |
| Forestry | 0.1 | -3.6 | 1.1 | 2.2 |
| Construction industry | 5.1 | 1.3 | 8.4 | 5.6 |
| Transportation | 6.6 | 7.6 | 6.9 | 5.4 |
| Trade and catering | 7.4 | 8.5 | 8.3 | 5.4 |
| Other production activities | 6.4 | 6.1 | 6.3 | 6.7 |

Source: *Trideset godina smoupravnog razvoja Jugoslavije 1950-80* (Belgrade: Savezni zavod za statistiku, 1981).

a slower pace, and its annual growth rate has averaged 3.8 percent. This situation has resulted in great changes in the structure of the Yugoslav economy. The industrial share of the total social product in 1980 increased to 40 percent, while the agricultural share decreased to 15 percent.

The highest growth rates were attained in the first decade of self-managed development (1950-60). Among the various economic sectors, the highest growth rate was achieved in the industrial sector. The data on the growth rate of the social product by decades and by economic activities are represented in Table 7.1.

It should be emphasized that the social product per capita in 1980 was 4.4 times greater than in 1950, while the social product in industry (per employed person) increased 3.6 times in the same period.

In the industrial sphere, dynamic development has been achieved and a series of new products has been introduced. The means of production, such as machinery, has been growing faster than the total production. The production of consumer goods has been growing

TABLE 7.2

Indexes of Development: Industry and Average Annual Growth Rates, 1950-80

| Physical Volume of Production | Indexes (1950 = 100) | | | Average Annual Growth Rates (in percent) | | | |
|---|---|---|---|---|---|---|---|
| | 1960 | 1970 | 1980 | 1951-80 | 1951-60 | 1961-70 | 1971-80 |
| Total | 262 | 582 | 1,186 | 8.6 | 10.1 | 8.3 | 7.4 |
| Means of work | 365 | 765 | 1,778 | 10.1 | 13.8 | 7.7 | 8.8 |
| Material for production | 254 | 552 | 1,038 | 8.1 | 9.8 | 8.1 | 6.5 |
| Consumer goods | 246 | 587 | 1,198 | 8.6 | 9.4 | 9.1 | 7.4 |
| Workers[a] | 197 | 257 | 369 | 4.8 | 9.9 | 2.7 | 3.7 |
| Labor productivity[a] | 139 | 238 | 337 | 4.4 | 4.2 | 5.5 | 3.6 |
| Value of fixed assets (1972 prices)[ab] | 213 | 228 | 708 | 7.8 | 9.9 | 8.5 | 8.3 |
| Power of running machines[cd] | 254 | 575 | 1,042 | 8.7 | 10.9 | 8.5 | 6.8 |
| Electroenergy consumption[a] | 517 | 1,050 | 1,731 | 10.7 | 22.8 | 13.3 | 5.5 |
| Consumption of imported raw materials[a] | 181 | 615 | 1,057 | 9.1 | 8.9 | 13.0 | 5.6 |

[a]1950 = 100.
[b]Year ending 1978.
[c]1951 = 100.
[d]Year ending 1979.

somewhat faster than the production of materials for reproduction. The rise of employment and labor productivity has occurred more or less at the same pace, but the energy needed for running machines has been rising at a faster rate than either employment or labor productivity and, in fact, faster than the total volume of production. All these indicators have been showing a downturn during the last decade (1970-80), as is shown in Table 7.2.

## INVESTMENT IN YUGOSLAVIA

Such dynamic development and intensive structural changes have been primarily owing to the high investment share in the social product. In the period from 1950 to 1980, the share of investment in fixed assets in the social product averaged 31 percent, while the average growth rate of investment was 8.3 percent.

It is also important to mention that the data on technical structures of investment indicate that the equipment share of total investment is constantly rising, while the share of the construction itself is decreasing. For instance, during the first phases of building the economy (1952), investment in civil engineering activities was large, with the share of construction activities in total investment amounting to about 56 percent, while the share of equipment was only 38 percent. This relationship between the investment in construction and equipment was reversed in 1978 and again in 1979.

It is important to consider the investment in new facilities and the investment in reconstruction as well. During the first phase of socialist building, the investment in new construction was dominant. Between 1975 and 1980, however, the following relations exist: the investment in new facilities is between 50 percent and 60 percent (30 percent to 40 percent of these are investments in new locations, and

TABLE 7.3

The Structure of Investment in Fixed Assets, at 1972 Prices (in percent)

| | 1952 | 1975 | 1979 |
|---|---|---|---|
| Economic activities | 84.0 | 70.5 | 72.0 |
| Industry and mining | 60.2 | 39.1 | 35.7 |
| Agriculture and fishing | 9.3 | 6.9 | 6.1 |
| Transportation and communication | 10.0 | 11.4 | 15.0 |
| Trade and catering | 0.8 | 5.8 | 6.5 |
| Other | 3.6 | 7.3 | 8.7 |
| Noneconomic activities | 16.0 | 29.5 | 28.0 |

TABLE 7.4

The Structure of the Social Product, at 1972 Prices
(in percent)

| | 1950 | 1960 | 1970 | 1975 | 1980 |
|---|---|---|---|---|---|
| Social product | 100.0 | 100.0 | 100.0 | 100.0 | 100.0 |
| Social sector | 70.6 | 72.8 | 81.5 | 83.0 | 86.7 |
| Individual sector | 29.4 | 27.2 | 18.5 | 16.1 | 13.3 |
| Industry | 21.5 | 28.4 | 33.6 | 37.0 | 39.2 |
| Agriculture and water management | 30.0 | 28.5 | 18.0 | 15.9 | 13.4 |
| Other production activities | 48.5 | 43.1 | 48.4 | 47.1 | 47.4 |

20 percent are investments in old locations), and the rest is reconstruction and modernization of the existing facilities.

The restructuring of the socioeconomic infrastructure to aid in the economic development of Yugoslavia has resulted in industry growing at a significant rate, as is shown in Table 7.4.

We can add to this the fact that the development of all regions of the country is of great policy importance and, as such, adds an additional dynamic and complex term to the Yugoslav development equation. It should also be noted that the battle against bureaucracy, the strengthening role of the working class, and the development of sound managerial practices are also significant contributors to the successful social development of Yugoslavia.

## ESTIMATION OF THE CONTRIBUTION OF TECHNOLOGY TO THE ECONOMIC DEVELOPMENT OF YUGOSLAVIA

Owing to the works of J. Tinbergen, R. Solow, and others, we have various models in which the social product is treated as a function of production assets, employment, and technical progress.

It should be noted that technical progress represents the rise in the social product that cannot be explained simply by the rise in fixed assets and the rise in employment. The social product can be increased by more efficient use of production assets without actually increasing the number of persons employed. The growth of the social product (Y) can be expressed as a function of the volume of production assets (K), the total number of persons employed (L), and time (t).

$$Y = f(K, L, t)$$

Dependence of the social product on time (t) expresses its dependence on technical progress.

By comparing the rate of growth of fixed assets (which was 8 percent in the period 1952-78) and the average rate of growth of the social product (which was 7.6 percent for the same period), we find that the efficiency of the utilization of fixed assets is decreasing. The growth rate of the social product (7.6 percent) consists of the percentage rise in employment (4.3 percent) and the percentage rise in labor productivity (3.3 percent).

We have taken into consideration a dynamic form of Tinbergen's production function[2] and the available data for the period from 1952 to 1978 to estimate contributions of fixed assets, employment, and technology to the rise of the social product in the Yugoslav economy.[3] The data are given in Table 7.5.

The weighted average of annual growth rates of fixed assets during the observed period is 8.17 percent ($= r_K$), and the same indicator for employment is 4.54 percent ($= r_L$), as calculated from the data in Table 7.5. Estimated coefficients $\alpha$, $\beta$, and $\gamma$ by least square method are: $\alpha = 0.577$, $\beta = 0.784$, and $\gamma = -0.305$. By putting these results in the equation we get

$$r_Y = 0.577(8.17) + 0.784(4.54) - 3.05 = 7.97$$

The result obtained is equal to the social product growth rate calculated as a mean value of annual growth rates for the observed period. (The mean value is 7.97 percent).

The first point that can be noticed is that the growth rate of technical progress is a negative quantity (-0.3 percent). But, the sum of coefficients $\alpha$ and $\beta$ is equal to 1.361, not 1. Since the technical progress is an expression of the increasing efficiency of fixed assets (where $\alpha + \beta > 1$), one can say that technical progress in our result is expressed by a negative growth rate (-0.305) and by the extent to which the sum of the coefficients $\alpha$ and $\beta$ is greater than 1 (1.361 - 1 = 0.361). However, if we apply to the estimated coefficients $\alpha$ and $\beta$ the classical postulate in which their sum has to be equal to 1, we will get new values for those coefficients.

$$\alpha' = \frac{\alpha}{\alpha + \beta} = \frac{0.577}{1.361} = 0.42$$

TABLE 7.5

Social Product, Fixed Assets, and Employment in the Yugoslav Economy

| Year | Social Product (Y) (in millions of dinars) | Employment (L) (in thousands of persons) | Basic Production Assets (K) | Annual Growth Rates (in percent) | | |
|---|---|---|---|---|---|---|
| | | | | Y | L | K |
| 1952 | 41,504 | 1,385 | 41,504 | — | — | — |
| 1953 | 44,033 | 1,471 | 44,033 | 6.1 | 6.2 | 6.7 |
| 1954 | 46,993 | 1,626 | 46,993 | 6.7 | 10.5 | 4.5 |
| 1955 | 52,083 | 1,799 | 52,083 | 10.8 | 10.6 | 9.4 |
| 1956 | 51,555 | 1,778 | 51,555 | -1.0 | -1.2 | -0.4 |
| 1957 | 60,269 | 1,917 | 60,269 | 16.9 | 7.8 | 9.6 |
| 1958 | 65,338 | 2,046 | 65,338 | 8.4 | 6.7 | 9.5 |
| 1959 | 75,092 | 2,266 | 75,092 | 14.9 | 10.8 | 5.9 |
| 1960 | 86,066 | 2,415 | 86,066 | 14.6 | 6.6 | 8.2 |
| 1961 | 92,772 | 2,637 | 92,772 | 9.0 | 9.2 | 9.7 |
| 1962 | 99,044 | 2,700 | 99,044 | 5.6 | 2.4 | 10.8 |
| 1963 | 113,018 | 2,764 | 113,018 | 14.1 | 2.4 | 9.1 |
| 1964 | 127,915 | 2,953 | 127,915 | 13.2 | 6.8 | 7.7 |
| 1965 | 132,475 | 3,009 | 132,475 | 3.6 | 1.9 | 8.7 |
| 1966 | 139,532 | 2,919 | 139,532 | 5.3 | -3.0 | 7.9 |
| 1967 | 142,820 | 2,894 | 142,820 | 2.4 | -0.9 | 6.9 |
| 1968 | 150,338 | 2,900 | 150,338 | 5.3 | 0.2 | 7.1 |
| 1969 | 164,808 | 3,011 | 164,808 | 9.6 | 3.8 | 9.2 |
| 1970 | 177,731 | 3,129 | 177,731 | 7.8 | 3.9 | 8.1 |
| 1971 | 193,863 | 3,279 | 193,863 | 9.1 | 4.5 | 10.0 |
| 1972 | 203,533 | 3,422 | 203,533 | 5.0 | 4.4 | 7.4 |
| 1973 | 212,250 | 3,495 | 212,250 | 4.3 | 2.1 | 7.5 |
| 1974 | 232,380 | 3,670 | 232,380 | 9.5 | 5.0 | 7.0 |
| 1975 | 243,179 | 3,877 | 243,179 | 4.6 | 5.6 | 8.0 |
| 1976 | 252,333 | 4,033 | 252,333 | 3.8 | 3.2 | 8.0 |
| 1977 | 274,090 | 4,182 | 274,090 | 8.6 | 4.5 | 8.3 |
| 1978 | 298,909 | 4,364 | 298,909 | 9.1 | 4.4 | 7.9 |

and $\beta' = \frac{\beta}{\alpha + \beta} = \frac{0.784}{1.361} = 0.58$

and $\gamma = r_Y - (\gamma' r_K + \beta' r_L) = 1.91$

The growth rate of the social product will then consist of the following:

$$r_Y = \alpha' r_K + \beta' r_L + \gamma' = 3.43 + 2.63 + 1.91$$

The relative contributions of the growths of basic production assets, employment, and technical progress to the growth rate of the social product are expressed in the following:

$$\frac{\alpha' r_K}{r_Y} + \frac{\beta' r_L}{r_Y} + \frac{\gamma'}{r_Y} \text{ or } 100 = 43 + 33 + 24 = 100$$

In conclusion, it can be said that the contribution of fixed assets to the rise of the social product was 43 percent in the period 1952-78, while the contributions of employment and technology were 33 percent and 24 percent, respectively.

It is clear from the discussion that technical progress manifests itself through new fixed assets and in turn demands greater education, skill, and productivity of the workers. That is the reason why production functions have to include materialized (nonneutral) technical progress. Therefore, the social product has to be treated as being a result of the rise of fixed assets and employment, materialized (nonneutral) technical progress, and neutral technical progress.

Technical progress manifests itself through new facilities and new equipment, as well as through a qualified labor force. That is why the increase of fixed assets and the increase of employment, together with other factors denoted as technical progress, represent complementary generators of the rise of the social product and of the rise in labor productivity. But the exact measurement of technical progress is still very questionable. Arnold Meertje expresses this position in the following:

> However, the problem of utilization of statistical and econometric methods, uncertainties in specification of various models, as well as an interpretation of results of empirical research, indicate that our knowledge concerning relations between growth, technical development and the factors of production is very far from being perfect.[4]

## DEVELOPMENT OF INDUSTRIAL PRODUCTION

Having in mind all that has been mentioned above, a presentation of additional empirical research and analysis of the Yugoslav economy can be illuminating.

The development of industry, measured either by the rise of investment or by the rise of the social product, has not been uniform during the past 30 years. The factors that have contributed to such development are the rise of labor productivity and the rise of employment. Of the total increase of fixed assets, 50 percent has contributed to the creation of new jobs; only 50 percent, then, has gone for the rise of labor productivity on the basis of a greater capital-labor ratio. Such a combination of development factors corresponds to the socioeconomic structure of Yugoslavia, since a large number of people have been employed or semiemployed in agriculture, and the birth rates have been very high. Both of these factors have been influencing the creation of available labor forces for industry.

To illustrate the dynamics of the means of production, consider the following: Yugoslavia was producing 18,026 tons of machinery for industry in 1950 and 137,375 tons in 1980; 1,388 tons of machinery for civil engineering in 1950 and 71,392 tons in 1980; 12,147 tons of agricultural machines in 1950 and 94,742 tons in 1980. About 60 percent of the need for equipment is being met through domestic production; the rest of the equipment needed is being imported. Yugoslavia was producing 2,408 million watts of electricity in 1950 and 59,435 million watts in 1980. The production of consumer goods has increased considerably, too. For example, 85,000 tons of sugar was produced in 1950 and 229,000 tons in 1980; 8,154,000 pairs of leather shoes in 1950 and 69,931,000 pairs in 1980; 19,000 tons of soap and washing powder in 1950 and 266,000 tons in 1980; electric ranges, washing machines, and television sets were not being manufactured in 1950, but by 1980 the country produced 640,000 electric ranges, 440,000 washing machines, 680,000 refrigerators, and 543,000 television sets.

During the last five years, the share of new industrial production in the total industrial production has been between 2.5 percent and 4.4 percent.

Technical progress is often expressed in terms of the introduction of new products or changes in existing patterns of production. From available data one can see that between 2,000 and 3,000 new products are being introduced annually in Yugoslavia. This is illustrated in Table 7.6.

During the observed period, significant changes in the economic system of the Socialist Federal Republic of Yugoslavia have occurred. These changes are also owing to the workers' right to organize them-

TABLE 7.6

New Industrial Production and Changes of Existing Products

| Year | New Products* Total | New Products* Actual | Existing Products Considerably Changed | Existing Products With Small Changes | Abandoned Products |
|---|---|---|---|---|---|
| 1967 | 5,737 | 1,026 | 4,949 | 9,206 | 10,575 |
| 1968 | 5,994 | 1,317 | 4,308 | 11,628 | 6,548 |
| 1969 | 6,125 | 2,143 | 3,055 | 14,495 | 7,233 |
| 1970 | 6,753 | 1,951 | 3,891 | 13,818 | 6,645 |
| 1971 | 7,520 | 2,307 | 4,838 | 14,855 | 7,177 |
| 1972 | 8,776 | 2,700 | 4,229 | 16,482 | 7,822 |
| 1973 | 9,373 | 3,173 | 4,884 | 17,076 | 7,037 |
| 1974 | 9,193 | 2,510 | 4,068 | 22,001 | 4,984 |
| 1975 | 10,442 | 3,648 | 3,055 | 17,803 | 4,533 |
| 1976 | 8,267 | 2,123 | 4,115 | 15,789 | 4,738 |

*"Total" stands for the sum of new products introduced by organizations of associated labor, while "actual" stands for the new products in real terms—that is, the duplication resulting from several firms introducing the same new product has been eliminated and the product is counted as a single product.

Source: Statistika dokument acija br. 4 (Belgrade: Savezni zavod za statistika, 1978).

selves into organizations of associated labor, expend their earned income as desired, enjoy the results of their work, and regulate their mutual relations.

## TECHNICAL EQUIPMENT OF LABOR (CAPITAL-LABOR RATIO)

The economic development of Yugoslavia owes a lot to industry, science, and technical progress. Besides the data on capital-labor ratio, measured by the increase in the value of equipment per worker and by the share of equipment in the total value of fixed assets, the data on electricity consumption, automation of working tools, and the

TABLE 7.7

Automation of Working Tools, 1975
(in percent)

| Industry | Total | Cycle Automatics | Automatic Transport and Production Multiplants | Semi-automatics | Other Automatics | Machines and Mechanized Tool Driving |
|---|---|---|---|---|---|---|
| Total | 100 | 13.1 | 14.2 | 31.6 | 6.1 | 35.0 |
| Energy supply/ energetics | 100 | 7.8 | 56.4 | 10.9 | 4.7 | 20.2 |
| Metallurgy | 100 | 21.7 | 25.0 | 25.5 | 5.4 | 22.4 |
| Nonmetals and construction materials | 100 | 21.5 | 2.1 | 36.0 | 4.9 | 35.5 |
| Manufacturing of metals, electro, and shipbuilding | 100 | 8.7 | 1.8 | 41.4 | 2.8 | 45.3 |
| Chemical and paper | 100 | 11.1 | 1.4 | 56.8 | 15.7 | 15.0 |
| Wood | 100 | 7.4 | 2.1 | 31.2 | 4.6 | 54.7 |
| Textile, leather, and rubber | 100 | 13.2 | 0.3 | 21.2 | 2.1 | 63.2 |
| Food and tobacco | 100 | 19.5 | 0.9 | 41.4 | 13.5 | 24.7 |
| Other | 100 | 24.9 | 0.1 | 37.3 | 6.6 | 31.1 |

TABLE 7.8

Fixed Assets and Equipment per Worker in the Social Sector of the Economy, at 1972 Prices
(in thousands of dinars)

| | 1952 | 1960 | 1970 | 1978 |
|---|---|---|---|---|
| Fixed assets | | | | |
| Total | 76.6 | 79.9 | 130.5 | 175.2 |
| Industry and mining | 85.3 | 95.5 | 148.3 | 204.1 |
| Agriculture with water | 31.4 | 43.0 | 144.1 | 261.5 |
| Forestry | 21.3 | 40.7 | 83.3 | 130.8 |
| Civil engineering | 11.1 | 15.8 | 31.0 | 50.8 |
| Transport and communication | 274.6 | 269.3 | 323.5 | 382.8 |
| Trade | 26.6 | 28.3 | 63.1 | 80.8 |
| Catering and tourism | 33.6 | 44.3 | 149.6 | 187.8 |
| Crafts industry | 6.7 | 8.3 | 63.3 | 56.1 |
| Production services and other production activities | 40.5 | 44.1 | 63.3 | 78.1 |
| Equipment | | | | |
| Total | 25.2 | 30.7 | 55.1 | 83.5 |
| Industry and mining | 40.1 | 43.0 | 68.6 | 103.5 |
| Agriculture and water | 9.7 | 18.9 | 43.2 | 90.0 |
| Forestry | 2.6 | 4.2 | 13.5 | 28.4 |
| Civil engineering | 7.9 | 12.1 | 23.0 | 39.2 |
| Transport and communication | 51.9 | 70.9 | 127.5 | 189.5 |
| Trade | 5.9 | 8.7 | 19.8 | 29.0 |
| Catering and tourism | 6.7 | 11.8 | 32.0 | 38.0 |
| Crafts industry | 2.9 | 3.7 | 31.3 | 28.8 |
| Production services and other production activities | 9.7 | 10.5 | 21.9 | 32.0 |

mechanization of agriculture are also interesting. The total consumption of electricity has increased by 25 times during the period 1950–80. It is interesting to note that the share of household consumption of electricity has also risen faster than the total consumption.

Another very important indicator of technical progress is the level of automation of working tools. The data by individual activities in this regard are given in Table 7.7. The data on values of fixed assets and equipment per worker by individual economic activities are given in Table 7.8.

TABLE 7.9

Technical Equipment of Labor and the Efficiency of Assets in the Social Sector, at 1972 Prices

| | 1952 | 1970 | 1975 | 1978 |
|---|---|---|---|---|
| Level of labor productivity[a] (thousands of dinars) | 29.9 | 57.3 | 63.5 | 69.0 |
| Technical equipment of labor[b] (capital-labor ratio) (thousands of dinars) | 25.2 | 55.1 | 72.2 | 83.5 |
| Efficiency of fixed assets[c] | 0.43 | 0.44 | 0.41 | 0.40 |

[a]Social product per worker in the social sector of the economy.

[b]Value of equipment per worker in the social sector of the economy.

[c]Ratio of social product to the value of fixed assets.

If one analyzes labor productivity and technical equipment of labor on the basis of trends in social product per worker and the efficiency of assets and equipment, one will see that, while labor productivity has been rising, the efficiency of equipment and fixed assets has been low. The reason for this situation lies in the fact that the technical equipment of labor has been greater than the rise of labor productivity. Such a situation is not only the consequence of insufficient increases in labor productivity but also of the faster development of capital-intensive economic branches (for example, electricity generators). In the case of Yugoslavia, it is also owing to the relocation of labor from agriculture to industry.

Technical equipment of labor and the efficiency of assets in the social sector of the economy are given in Table 7.9.

## DEVELOPMENT OF SCIENCE AND EDUCATION

The development of Yugoslavia (and its transformation from an agricultural to an industrial country) is a result in large measure of the development of education, scientific and research work, and scientific institutions.

Today 99 percent of children under age 10 receive primary education, and 92 percent of children between ages 11 and 14 receive

TABLE 7.10

Scientific and Research Organizations

| | 1965 | 1970 | 1975 |
|---|---|---|---|
| Total scientific and research organizations | 432 | 463 | 501 |
| Scientific and research cadre | 8,380 | 10,443 | 12,854 |
| Finished works | 5,575 | 9,789 | 12,961 |
| Independent scientific and research organizations | 276 | 261 | 270 |
| Scientific and research cadre | 6,981 | 7,120 | 7,901 |
| Finished works | 3,960 | 6,786 | 7,699 |
| Scientific units within academies of science | 36 | 29 | 36 |
| Scientific and research cadre | 107 | 113 | 215 |
| Finished works | 81 | 119 | 416 |
| Research and development units within work organizations | 120 | 173 | 195 |
| Scientific and research cadre | 1,292 | 3,210 | 4,738 |
| Finished works | 1,534 | 2,884 | 4,846 |

secondary/high school education. Thus, the illiteracy rate has fallen to 10 percent of the total population. There is practically no illiteracy in the segment of the population under age 35. In 1979 there were 54,178 graduate students; 220,325 pupils finished secondary school; 1,579 persons obtained a master of science degree; and 846 persons received their Ph.D.'s.

The number of scientific and research organizations and their achievements appear in Table 7.10. Table 7.11 gives the breakdown by fields of specialization.

Technical-scientific progress has also come from the acquisition of rights for the utilization of foreign technology through licenses. Over 90 percent of the licenses purchased from foreign countries are being used in industry. About 40 percent of the licenses used in industry are being used in the manufacture of chemical products, 15 percent in metal manufacturing industries, 13 percent for the production of transportation products, and 8 percent for electromachines and reparations. In 1979, 465 contracts to buy licenses were concluded. In the same year, 14 domestic licenses were sold abroad.

The transfer of techniques and technology that influences the development of industry is carried out by joint ventures. In 1979 there

TABLE 7.11

Scientific and Research Organizations by Scientific Fields, 1977

| | Organi-zations | Scientific Workers (researchers and profes-sional and technical workers with university education) | Finished Research Works |
|---|---|---|---|
| Total scientific and research organizations | 439 | 12,369 | 11,567 |
| Natural/mathematical sciences | 57 | 1,580 | 1,248 |
| Technical sciences | 169 | 5,356 | 6,798 |
| Medical sciences | 27 | 1,173 | 210 |
| Agricultural/forestry sciences | 66 | 1,578 | 1,019 |
| Social sciences | 120 | 2,681 | 2,292 |
| Total research and development organizations within organizations of associated labor | 136 | 3,095 | 2,427 |
| Natural/mathematical sciences | 7 | 298 | 123 |
| Technical sciences | 88 | 1,856 | 1,575 |
| Medical sciences | 8 | 295 | 308 |
| Agricultural/forestry sciences | 23 | 462 | 283 |
| Social sciences | 10 | 184 | 138 |

Source: Statisticki godisnjak Jugoslavije, 1979.

were 128 joint ventures involving resources for investment and modern technology, enabling labor productivity to improve. There is also significant cooperation between Yugoslav institutions of science and education and corresponding institutions elsewhere in the world. Yugoslavia is a country that has close relations and cooperation with many countries on a bilateral and multilateral basis, as well as through the United Nations.

## IMPLICATIONS FOR THE FUTURE

Economic development is a constant process in which technology and technical progress play decisive roles. That law of the contemporary world has been manifested in the economic development of Yugoslavia, as well as in many other countries. One can ask the question, Will technology have the same role in the coming 20 to 30 years as it has had in the previous 30? The investigations dealing with this problem prove that the technology in the period to come will have the most significance in economic development. This can be particularly applied to the developing countries in which numerous technological accomplishments and results in development have not been used but have been implemented in developed countries. Yugoslavia states that the strategy of its economic development will also depend in the future on the strategy of technological development described earlier. The processes of change in the structure of the economy, the need for further improvement of work productivity, and the rising standard of living impose the need for even faster development and use of technology. Politically, the preservation of its economic independence internally and nonalignment in foreign policy also require faster development of science and technology and their application to the development of Yugoslavia. Bearing these points in mind, this chapter has tried to present a framework for the assessment of technology in the future development of Yugoslavia.

## ANALYSIS AND STRATEGY

Before establishing the basis of the strategy and policy for technological development, it is important to analyze the experiences of the past and achieved goals and conditions. The analyses establish that the country has, in fact, attained considerable progress in the areas leading to the development of new technology. At the same time, it has been recognized that the progress achieved in the educational process, as well as the program orientations of the scientific institutions, were not sufficiently connected with the needs and trends of the country. Both education and scientific work are burdened with the old classic opinion of separation between the organization and practice. Serious analysis also reveals the inadequacies in the structure of scientific institutions, while certain indigenous capabilities have also been created.

The importation of technology, acquisition of licenses, and industrial cooperation and other forms of technology transfer, no matter how necessary, tend to result in the neglect of innovative activity and technological progress within the country. Foreign technologies have enabled the owners of these technologies, particularly transnational companies, to dictate the conditions of the transfer of technology, thus adversely affecting the economic development of the country. The analysis of agreements on business-technical cooperation between Yugoslavia and some transnational companies in certain production branches (such as the chemical industry and pharmaceuticals) proves that getting a new technology is only a partial solution, which usually results in a dependency on foreign technology that, more often than not, is obsolete.

When producing the strategy of technological development in general, but also in specific sectors and branches of the economy, Yugoslavia faced several dilemmas and questions dealing with the degree to which development should be based on domestic and imported technology. These considerations also apply to technology imported from other developing countries. In setting the strategy for the technological development of Yugoslavia, the more dynamic sectors had to be given priority, since these sections produce the most. In the next phase, it is expected that technological changes will be made in the following sectors: electronics, biotechnologies, energy technologies, the mining industry, basic iron and steel and nonferrous metallurgy, and automation of production and management. New development should also be expected in agricultural production and forestry.

The strategy for technological development in Yugoslavia is also concerned with the introduction of work and capital-intensive technology. A precise analysis of this problem within particular branches showed that both labor-intensive and capital-intensive technologies are needed while closely monitoring and evaluating the social impact brought about by these introductions. For instance, when erecting electric power plants, the production capacities are capital intensive, while the work on a high-voltage feeder or new lines will be labor intensive. It is similar in the metal sector, where steel production will be capital intensive, and the processing of metal and production of engines will be labor intensive. Any assessment of technologies must address the issues and problems of increasing possibilities of employment in the tertiary sector and in developing cooperation between large and small firms, as well as between the state and private sectors. An examination of the agreements on long-term industrial cooperation with foreign firms covering the period between 1968 and 1977 reveals that, of 492 contracts, 60.9 percent were technological capital-intensive contracts, 22.5 percent labor-intensive, and 16.6 percent combined contracts.

Closely connected with the strategy of technological development is the more direct combination of that strategy with overall economic development targets, the use of domestic resources, and an improvement in the efficiency and quality of business management. The task was to find out how much could be achieved using the new domestic technology and how much using the foreign, as well as to identify the sectors and technological areas in which the development of appropriate science will play crucial roles in the formation of domestic scientific research institutions.

In summary, the new strategy of the technological development of Yugoslavia has been assessed as follows:

1. Accelerate development and strengthen scientific research and technological applications in the country.
2. Transfer appropriate technologies from developed countries through the various cooperative channels.
3. Widen the scientific-technical cooperation with developing countries, both in terms of importing technologies and providing technological support and assistance to those in need.

The strengthening of scientific institutions and scientific research work, including the development of new technology and innovations, is one of the basic choices in Yugoslavia's future strategy for development. Greater financial resources will be spent in the future on basic research instead of the current expenditure of one percent of the national income. Additional resources will be used to promote scientific and technological activities both in government bodies and in self-managing communities. The money will also be used to set up new research institutions and to promote scientific research work in firms. There are plans to intensify the cooperation of universities and faculties and particular national planning units and organizations. State policy will be to stimulate firms by taxation exemption where they make investments in scientific research work, encourage innovations, and facilitate the application of domestic innovations in practice.

Yugoslavia's strategy for technological development is heavily dependent upon technology transfer and cooperation from the developed countries. Therefore, efforts will be made to resolve any area of conflict or negative effect of the previous forms of cooperation and to emphasize the positive aspects of such cooperation. Joint ventures and technical cooperations are essentially better when compared with the purchase of licenses and patent rights.

The five-year plan of development for Yugoslavia for the period 1981-85 forsees a fairly dynamic development growth and a GNP of 4 percent to 4.5 percent per annum. This growth will be achieved through the use of existing resources, resulting in an increase in the number of employed and a considerable increase in work productivity. Here, special importance has been given to the three sectors within which science and technology are expected to make special contributions. These areas are energy development and better use of energy resources; agricultural improvements leading to higher yields; and business and industrial improvements leading to a greater capacity for production, export earnings, and standards of living.

## CONCLUSION

In creating strategies for the technological development of Yugoslavia, the future focus will be directed more toward cooperation with the developing countries. This policy serves the intent of nonalignment and widens and improves economic and other relations with these countries. In order to realize this goal, it is necessary to encourage contacts with scientific institutions and other development organizations in developing countries. It is also imperative that Yugoslavia develop background information on the state of the art of science and technology in these countries, as well as on their cultures, in order to facilitate appropriate cooperation and exchanges of expertise.

The analysis of the development of Yugoslavia offers the following insights: technology, fixed assets, and employment are three factors that have given significant impetus to the successful development of the country. A considerable volume of investment, as well as the structure of the investment itself, has paved the way not only for successful economic growth but also for important structural changes. These changes have had an impact on technical and technological progress, on the introduction of new products, on the equipment of labor, and on the rise of labor productivity. All these are the result of the transfer of technology from developed countries, increased education, and development of the country's own scientific and research cadre, as well as of the planned market model of the self-managing economy of Yugoslavia.

## NOTES

1. Social product is a term mainly used in Yugoslavia to denote national income plus depreciation. This indicator is comparable to GNP in Western countries.

2. If we have a production function of form $Y = AK^{\alpha}L^{\beta}e^{\gamma t}$, then the equation that expresses the growth rates of the observed variables from the above production function

$$r_Y = \alpha r_K + \beta r_L + \gamma dt$$

is called Tinbergen's dynamic production function, where symbol r = growth rate of corresponding variable.

3. The equation (x) represents a linear regression model, with unknown parameters $\alpha$, $\beta$, and $\gamma$. Growth rates of social product ($r_Y$), employment ($r_L$), and fixed assets ($r_K$) are known (weighted averages of corresponding growth rates for an observed period). The estimation of unknown parameters can be done by least square method.

4. Arnold Meertje, Economics and Technical Change (London, 1977), p. 202.

# 8

# SCIENTIFIC AND TECHNOLOGICAL REVOLUTION AND THE LABOR OF WOMEN IN THE USSR

VALENTINA NIKOLAYEVA TERESHKOVA

The last quarter of the twentieth century is bound to go down in world history as an epoch of scientific and technological breakthroughs (and their wide application in practice) and as an epoch of a scientific and technological revolution of great historic importance and scale.

Certainly, one cannot ignore the darker sides of scientific and technological progress primarily related to industry-induced pollution of the environment. But this pollution is far from inevitable. Humanity knows effective, if costly, antipollution techniques—purification technology, but most important, closed technological cycles. A broad application of these and other antipollution techniques calls for stable and peaceful conditions for the development of our planet, conditions that would make it possible to use humanity's material and intellectual resources for constructive not destructive purposes.

Experts agree that only by spreading advanced technology will it be possible to considerably help meet many current needs to the greatest possible degree. The food problem, environmental protection, access to needed energy and mineral resources, demographic aspects of development, and so on are all complicated global problems that await solution.

For young nations to overcome the backwardness they inherited from the time of colonial oppression, to exercise full sovereignty over their natural wealth, to attain true political and economic independence, and to bring about a new world economic order by way of restructuring international economic relations on a democratic and equal footing—doesn't this body of problems constitute another pressing task of our times? Indeed, it is primarily a social task, but it also has aspects related to the positive use of scientific and technological advances for the betterment of humanity.

Finally, there is a problem whose solution is central to the very survival of the human race—the prevention of the military use of nuclear technology. This means preventing a nuclear war and saving humanity from the threat of annihilation. This is not only a precondition for the further development of the community of people, but an objective condition for any social progress at all. No other problem is more acute or more pressing today.

The nature and scale of the above-mentioned problems are such that humanity cannot afford to waste a vast part of its economic, scientific, and technical potentials on the arms race. It is necessary to emphasize that, intractable as the problems facing humanity are, there exists in principle a possibility of solving them in a humanitarian way. In fact, this is not only a probability but an urgent need. Such is the call of the times.

Addressing the 26th Communist Party of the Soviet Union (CPSU) Congress, L. I. Brezhnev said, "If there is peace, the creative energy of the peoples backed by the achievements of science and technology is certain to solve problems that are now troubling people" —a reminder that peace is an essential condition for human progress.

It is hardly possible within the confines of this chapter to consider all the aspects of using modern technology for the good of women and children. This chapter will touch only on those aspects related to my own experience: the contribution of space research to current development and the real position of Soviet working women in the conditions of scientific and technological progress.

Hopefully, the Soviet experience of using scientific and technological advances in the interests of social development and the well-being of all Soviet citizens and the experience of the relationship between people and machinery under socialism will interest some readers.

What opportunities for solving the above problems are opened by one of the main achievements of this century's scientific and technological revolution—the breakaway of humanity from the earth's gravity and the exploration and practical use of infinite outer space for peaceful purposes?

Until quite recently, there were many arguments and doubts about the expediency of space research, since it consumes large amounts of material resources. These doubts have been dispelled by the scientific and practical results obtained in as little a time span as the last 20 years: the first manned space flight; extravehicular activity; the studies of the moon with automatic spacecraft; the flights of robot stations to planets of the solar system; the landing of these crafts on their surfaces; the transmission to earth of the images of their landscapes; the launching of sophisticated astronomical instruments into the earth's orbit; the landing of spacemen on the moon's surface; the establishment of habitable orbital stations supported by transport and cargo craft; the endurance missions on board; the establishment of a network of weather, navigation, and communication satellites; and the beginning of research in remote sensing and space technology.

The experience of our country, which has always urged using space for none other than peaceful purposes, shows that the most effective way of conducting long-term space research is to use long-service orbital lab stations with changeable crews. The on-board systems and research equipment on such stations are used for several missions. As a result, the research done on them costs less than that done aboard independent spacecraft that have similar equipment and the same mission.

Orbital stations open fundamentally new opportunities because they can carry large research instruments and can be used for research for long periods of time. Thanks to these advantages, the stations have proved very convenient for a wide range of research in various fields, such as long-term measurement of the current of micrometeors and the various elementary particles in space, as well as in various programs with the help of telescopes.

The registration of radiation from various parts of the earth's dry surfaces combined with visual observations by spacemen provides an opportunity to supply speedy and important information to geologists, hydrologists, meteorologists, and forestry and fisheries experts. What is also important is the development of space research methods to forecast crops, evaluate the state of pastures, determine water supply, identify water pollution, assess the damage caused by calamities, and tackle other vital problems.

The orbital stations are also convenient for technological research (to obtain substances with new properties in weightless conditions) and for medical and biological research (to develop technology for producing pure biological substances in zero gravity, vaccines, medicines, and ferments).

All this makes it possible to get a better insight into the processes that underlie the development of the universe, to come closer to a scientifically based theory of the emergence of life on earth, to study the relationship between the earth and the sun, and to better understand the earth itself. Space research means having a whole range of advantages over ground means. It makes it possible to conduct long-term global observation of the earth's surface in any weather conditions, including those regions that are hard or even impossible to get to. It is difficult to overestimate the importance of the means of space communication: satellites of the Molnya type ensure stable telephone and television communications. Satellites can also be used to convert solar energy into electricity and transmit it to earth.

Space research that serves the interests of all humanity may be helping in the comprehension of problems facing humanity as an integral whole more than any other achievement of the scientific and technological revolution.

When talking about space programs, one should mention the participation of women in them. In our country these programs from their very inception have been providing for the participation of women in various types of space research, as well as the theoretical education and practical training of cosmonauts. The process of selection to enter the programs, both for men and women, takes into account scientific and technological capabilities and training achievements of the participants.

To use the positive sides of the scientific and technological revolution to the fullest degree and neutralize its potentially negative effects—these are the goals the USSR is trying to reach domestically and internationally. In the USSR the scientific and technological revolution helps to solve social and economic problems. This was emphasized at the 26th CPSU Congress (March 1981), and was written into the Eleventh Five-Year Economic Development Plan. "The central objective of the Eleventh Five-Year Plan is to ensure the further improvement of the Soviet peoples' well-being on the basis of the steady and constructive development of the national economy, and accelerated scientific and technological progress."

The scientific and technological revolution in the USSR is a nationwide affair in which women take a most active part on an equal footing with men. The high labor activity of Soviet women is stipulated both by objective and subjective causes. The objective causes are the constantly growing needs of the national economy in the work force, including the women's work force, which is connected with the expansion of production and the creation of new jobs as a result of the mechanization and automation in production.

It should be noted that the participation of women in public production is increasing despite a steady growth in the living standards of the population. From 1940 to 1971, real incomes in the USSR tripled, and the share of women in production increased from 39 percent to 51 percent.[1] This fact shows, above all, that subjective causes have come to play a greater role. Among these, special mention should be made of the interest women have in social contacts, in developing their personality, and of their need for professional labor. It is professional labor that opens the way to women's equality in the family and society and constitutes the material stimulus to greater social activity by women. Professional activity is becoming an area for women's self-expression and self-assertion. At present, the USSR has practically ensured the full engagement of women in socially useful work. Over 90 percent of the nation's able-bodied women either work or study.[2]

Scientific and technological progress not only widens the sphere of women's work but also leads to the improvement of working women's skills. The representation of women is growing especially quickly in those professions and trades that are central to raising the technical level of public production and to upgrading its efficiency. Women account for 35 percent of the industrial designers and for over 45 percent of those employed in technological research.[3]

Scientific and technological progress has also caused changes in the pattern of women's employment in industry. More women (including engineers and technical personnel in managerial positions) have come to work in the machine-building, chemical, electronic and electrical and radio engineering industries (that is, industries that are in the forefront of technological progress). The wider application of computers has given rise to a new group of professions (programmers, coders, punch card machine operators, main and auxiliary machine operators, and mechanics). Women constitute between 50 percent and 60 percent of this personnel.[4]

In the USSR industry has now become the main area employing women's labor; it ranks first in the total number of women employed. Among the branches of industry, engineering holds first place in the number of women it employs (42 percent of the engineering industry's total work force in 1974).[5] At present, women experts employed in the economy account for 59 percent of all the experts with a higher and specialized secondary education.[6]

The growing number of women doing mechanized work both in industry and in farming is a trend that is steadily increasing under the influence of the scientific and technological revolution. As a result of the comprehensive mechanization and automation of farm operations and the transfer of cattle breeding to an industrial basis, the work of women farmers is changing in nature and content, and it is approaching ever more closely the work in industry. In farming the share of women experts is becoming greater, too. Women make up more than one-third of the agronomists, stockbreeders, and veterinary surgeons with a higher and specialized secondary education.[7]

A logical consequence of the radical changes that have come about in the content of work under the influence of the scientific and technological revolution is a further intellectualization of women's work. In the past 20 years, the number of women in science has tripled. Today, women comprise 40 percent of the nation's researchers.[8] Characteristically, women researchers are tackling, among other things, such problems as thermonuclear fusion, space biology, mathematical physics, genetics, high-molecular compound chemistry, radio engineering, minerology, and so on.

The current scientific and technological progress is making ever greater demands on the creative attitude to work. The USSR provides ample opportunities for innovative thought—the necessary condition for the development of the scientific and technological revolution. In 1980 there were 2.5 million women in the local branches of the All-

Union Society of Inventors and Innovators (35 percent of its membership).[9] The development of technical creativity by Soviet women reflects not only the growth of their initiative but satisfaction with work as a creative process as well.

It is known that the scientific and technological revolution necessitates higher cultural and professional standards from all those engaged in public production. These requirements are being met to an increasingly fuller extent by the Soviet educational system, which has undergone great changes in recent years.

The USSR has completed the switchover to a compulsory, full secondary education. Gratuitous education at all levels, instruction in the native language (which is especially important for the USSR, a multinational state), and the availability of the necessary number of daytime and night schools—all of these make it possible to put into effect Article 45 of the USSR Constitution on compulsory secondary education for the entire population.

Soviet women are on a par with men in terms of education, which at the same time is a testimony to, and one of the chief guarantees of, their equal position in society. Women constitute 52 percent of the students at institutions of higher learning, including 41 percent of the students of technical institutions of higher learning. The share of women in specialized secondary educational institutions is 56 percent.[10]

In the USSR women have equal rights to receive not only a general education but also vocational training at all levels. Upon graduating from school a girl can freely choose a trade or profession. She can continue her education at a vocational training school, a technical school, or a higher educational institution. Nearly one million skilled women workers are trained every year at 6,000 vocational schools. The number of trades open to girls has reached 1,000.[11]

It should be particularly noted that the system for planning the training of personnel in the USSR ensures jobs in line with the trade and skill obtained for all graduates from educational establishments (men and women). In guaranteeing women the right to work, the socialist economic system fully precludes any discrimination in employment and in pay. Under the conditions of the present-day scientific and technological revolution, primary training is only the first stage in educating a skilled worker. The knowledge mastered in an educational institution sometimes proves insufficient; such is the rate of improvement in machinery and technology.

In the USSR there is a ramified network of courses for upgrading the skill of both the workers and engineers, technical personnel, and other specialists. One can improve his or her skill on a full-time basis (while continuing to collect full or partial pay) or on a part-time basis (while staying on the job). There are many ways of improving one's skill directly on the job: factory-run vocational schools, experience-exchange schools, the learning of a second or borderline trade, and so on. Under the conditions of the scientific and technological revolution, training and retraining is of high priority in the Soviet educational system; this is of special importance to women.

Those workers, men and women, who study on a part-time basis at night and in correspondence technical schools, institutes, and schools for working youth in industry and farming, are granted large benefits by the state. They enjoy paid leaves to take examinations and favorable work schedules during the academic year.

The scientific and technological revolution in the USSR, as in other countries, results in the freeing of the work force. Not a single worker, however, whose trade has become redundant in connection with the introduction of a new technology is left without a job. Owing to the stable growth rate of socialist production, all workers, men and women alike, are guaranteed full-time employment.

The entire system of people's education, management studies, and on-the-job training and retraining is closely connected to the advancement of scientific and technological progress and operates in accordance with the plans drawn in advance. Thanks to planned economic development, the appropriate government bodies, enterprises, and trade unions know what changes will occur in machinery, technology, and the organization of production. This makes it possible to train, retrain, and improve workers' skills within the framework of the comprehensive social and economic development plans of enterprises. These plans envisage concrete measures to improve the housing conditions, health care, cultural services for the employees, and so on.

The plans of many enterprises also provide for the training of women and for upgrading their skills. The need for such special measures stems from the fact that working mothers are not always in a position to use training opportunities, though these are equal to those of men. Because of the greater workload in the home, they have less time to improve their skill. For this reason, the decision of the CPSU Central Committee and the USSR Council of Ministers "On Measures for Further Improvement of On-The-Job Training and for Skill Improvement" (June 1979), among other things, gives women with children under eight years of age an opportunity to upgrade their skill on a full-time basis while drawing average monthly wages.

The growing labor activity of women within the context of the scientific and technological revolution and the intensification of production means creating more favorable conditions for highly productive work, not an increase in the women's work load. To accomplish this, socialist society is trying to take into account as fully as possible the psychophysical characteristics of women workers. In this connection, it should be noted that Soviet society does not regard equality between men and women as an arithmetical equation.

In regarding women's equality in work as equality in getting a job, mastering and improving skills, promotion, pay, and so on, one also means the right of women to special work conditions and conditions in the home that are conducive to maternity and to the all-around development of personality. Women enjoy a number of privileges in the regime of their work: favorable terms of retirement, paid maternity leave, the right to return to the same job after childbirth, paid breaks at work to feed the child, special job safety during pregnancy, partially paid one-year nursing leave, and so on. The policy of the USSR has always been aimed at protecting women's labor, and the basic principle of this policy is to eliminate those factors in their working conditions that are hazardous to women's health.

The USSR has, in recent years, renewed the list of those trades and occupations that are closed to women because of their hard or hazardous conditions. The list was drawn up by research organizations of the USSR Ministry of Health, the USSR State Committee for Labor and Social Questions, the labor-protection institutes of the All-Union Central Council of Trade Unions, and research institutes of various industries. The main criterion for assessing whether a profession was a health hazard was the potential danger of an adverse influence on women's health, including their childbearing function and the health of posterity. The new list, totaling 460 trades and occupations, was enacted in 1979.

Displaced women workers were reemployed in keeping with their wishes and professions. Under this act, for example, women are not allowed to be operators of some technological stages of nonorganic production, but they can be operators in another stage with more favorable working conditions.

The USSR strictly observes UN Convention 45, which it ratified, on banning the employment of women in underground jobs. By decision of the Soviet government, women are not allowed to work underground in the mining industry or to build underground projects.

In the USSR a wide range of problems concerning the protection of women's labor is handled as early as in the designing stage. The specific features of the woman's body are considered during the reconstruction of existing industrial and farm equipment and the development of new types of this equipment.

Thus, by paying great attention to making women's work healthier, the USSR is taking practical steps to remove health hazards from women's working conditions.

There are 29 research institutes where experts on various subjects (doctors, physicists, chemists, sociologists, microbiologists, and designers) are working to improve labor protection. They focus on ensuring job safety, eliminating occupational diseases and accidents, and removing factors that are harmful to the health of women workers. On-the-job control of labor protection is exercised by trade unions through safety inspectors who are given special rights by law.

The new position and role of women in a socialist society gives rise to new problems, too. These are examined by governmental scientific and public organizations, and ways of solving them are written into state plans for the nation's social and economic development. Some of these measures include making social and everyday services a highly developed modern industry, further increasing the number of places at preschool children's institutions to develop various forms of out-of-school work with children, perfecting ways of training and retraining women and upgrading their skills, and so on. An important role in supervising the implementation of the state plans and laws belongs to the standing deputies' committees of the USSR Supreme Soviet and the Supreme Soviets of the Union and autonomous republics, of regions, and of territories; they consider questions of women's working and everyday life conditions and on mother and child care. Considering the extensive variety of social functions for women under the conditions of the scientific and technological revolution, the socialist state concentrates its efforts on creating the best possible conditions for women to combine their duties as mothers, workers, and citizens.

In the USSR the guiding principle for educating the younger generation is the unity of family and public efforts. The principle was developed on a scientific basis and has proved efficient. There is a ramified network of preschool children's institutions for working mothers; at present over 14 million children attend crèches and kindergartens a day. Five million children are taken care of at preschool institutions that operate on a seasonal basis. During the Eleventh Five-Year Plan (1981-85) the number of places at preschool children's institutions will grow by 2.5 million. (It should be noted that the upkeep of one child at a preschool children's institution costs the state R400-R500 a year, while the parents contribute no more than one quarter of that expense.)

Besides paying greater attention to the public forms of out-of-school work with children, much is being done to enhance the role of the family in the educational process. It is common knowledge that an important role in social progress and the spiritual education of the working people is played by leisure time. Naturally, leisure time increases through the reduction of working hours. The USSR has a 41-hour workweek. But since women do more work in the home, there is still a difference in the free time available to men and women. Yet, as production forces develop on the basis of the scientific and technological revolution, everyday services expand, more household appliances appear, and the traditional distribution of the workload in the family changes, the gap in free time available to men and women gradually grows narrower.

It is significant that in view of the social importance of women being able to combine production, social, and family activities, the question of reducing the gap in free time between men and women is given prominence in so fundamental a document as the CPSU Program, which sets the goal of wiping out the remaining inequality of women in the home.

On the subject of the steady growth of the Soviet people's material well-being, it seems of particular significance to note the role of public consumption funds (that is, that part of the national budget citizens receive in the form of free material benefits and services and in the form of various payments). The public consumption funds go to finance free education, free medical services, benefits (pensions and other kinds of social security insurance), paid leaves, student grants, free or partially paid for accommodation vouchers to sanatoria and holiday centers, and so on. Soviet society is known to be helping all of its members (whatever their social status, sex, nationality, or family income) to keep in good health, raise their educational standards, and satisfy various material and cultural needs.

If one addresses one's self to the Program of Action adopted by the World Conference of the UN Decade for Women (Copenhagen, July 1980) and also to the convention on the elimination of all forms of discrimination against women, which the USSR signed and ratified in 1980, one will see how much attention is given to problems related in some way or other to the scientific and technological revolution: integration of women into development, equality of women in education and training, the contribution of women to the struggle for peace and disarmament, and so on.

This is natural. The flow of time is impetuous, and how the people of the earth handle the problems of scientific, technological, and social progress on the threshold of the third millenium will determine the fate of our planet in the twenty-first century.

## NOTES

1. E. Novikova, V. S. Yazykova, and Z. A. Yankova, "Women, Family and Work" (Moscow, 1978).
2. "Women in a Socialist Society" (Moscow, 1980).
3. Novikova, Yazykova, and Yankova, "Women, Family and Work."
4. Ibid.
5. Ibid.

6. "Women in the USSR" (Moscow, 1981).
7. "Women in a Socialist Society."
8. "Women in the USSR."
9. T. N. Sidorova, "Work and Modern Woman" (Moscow, 1981).
10. "Women in the USSR."
11. "Women in a Socialist Society."

# 9

# TECHNOLOGY ASSESSMENT IN INDIAN AGRICULTURE

AVTAR S. ATWAL

Agriculture is the prime occupation of 72 percent of the 680 million people in India, and it accounts for 48 percent of the national gross income. During the 1950s and 1960s, India experienced great food shortages and had to import huge quantities of foodgrains, the maximum being 10.34 million tons in 1966. Subsequently, the country achieved not only self-sufficiency in foodgrain requirements but also gathered some surpluses and in 1979 even exported 0.9 million

TABLE 9.1

Pattern of Land Use in Punjab
(in thousands of acres)

| Year | Geographical Area | Net Area Sowed | Total Cropped Area | Cropping Intensity (percent) |
|---|---|---|---|---|
| 1960/61 | 12,449 | 9,283 | 11,699 | 125.9 |
| 1965/66 | 12,449 | 9,397 | 12,081 | 128.5 |
| 1970/71 | 12,449 | 10,015 | 14,031 | 140.1 |
| 1975/76 | 12,449 | 10,275 | 15,457 | 150.4 |
| 1979/80 | 12,449 | 10,329 | 16,151 | 156.4 |

Source: Punjab Government, Statistical Abstracts of Punjab (Chandigarh).

tons after meeting the requirements of a 2.5 percent annual increase in population.

In this study of Punjab agriculture, an assessment has been made of the adoption of agricultural technology during the last two decades in terms of cost-benefit ratio and other attributes. The state of Punjab, which has made the most rapid progress in agriculture in India, is located in the northern region bordering with Pakistan. It covers approximately 1.5 percent of the geographical area of India and constitutes about 2.5 percent of the cultivated area of the country. The land use pattern in the last two decades is shown in Table 9.1.

After the partition of India in 1947, most of the fertile canal-irrigated land went to Pakistan; what was left on the Indian side of Punjab was relatively poor land. The farmers were agriculturally backward and socially tradition bound. They were subjected to a vicious circle of poor resources, low production, and persistent indebtedness. Sometimes their ancestral debts were so heavy that they had to forfeit ownership of their land in exchange for the money they owed to the private moneylenders. Hence, litigation and other social problems were the direct outcome. Since the holdings were fragmented, there was no incentive for traditional farmers to install energized tube wells. Hence, they persistently bore the drudgery of lifting water with the ancient Persian wheel.

With partition of the country in 1947, 4.75 million people were uprooted in the combined Punjab. The Muslims from the Indian side went to Pakistan and the non-Muslims of West Pakistan came mostly to East Punjab. The population that came to India consisted of businessmen, shopkeepers, servicemen, and cultivators who were mostly landowners. The occupational differences between those who came and those who left Punjab created many problems in the socioeconomic matrix of the rural areas.

A sizable proportion of the farmers (cultivators) who came to Punjab were either exservicemen or the descendants of those who had established the Canal Colonies, Neelibar and Ganjibar, which they left behind in Pakistan. Apart from the fact that they got less land in Punjab, they also had to face physical, sociological, and psychological odds. The loss of property, the sense of displacement, and the lack of facilities for proper rehabilitation created much frustration among the refugees who had come on foot convoys and, hence, could not bring much with them. Since they were faced with the challenge of struggling for survival, they worked diligently and tirelessly. They also depended on private moneylenders for obtaining credit at high rates of interest. They worked desperately in order to begin a new life. The refugees eventually settled throughout Punjab, and their villages formed a mosaic among the existing old villages. They were allotted refugee lands in solid blocks; thus, they got a better start than the indigenous farmers, whose holdings were fragmented owing to generations of family divisions.

After the partition of India and Pakistan, another significant experiment was conducted through the establishment of Garden Colonies in selected villages well distributed in the State. Educated and comparatively more enlightened refugee farmers were given the choice of resettlement there. The idea was to plant orchards and raise high-value fruits. A good start was made, but the orchardists, after trying their hand at horticulture and failing, shifted to potato cultivation and vegetable growing. Later, they also played the most significant role in influencing the other farming communities around them and in bringing about the Green Revolution. Traditional bullock cultivation, poor irrigation facilities, growing of local varieties of crops, low application of fertilizers, heavy dependence on monsoons, and a poor market infrastructure without price supports were gradually replaced by tractor cultivation, lifting of groundwater by energized pumping sets and tube wells, the growing of dwarf Mexican wheat varieties and International Rice Research Institute (IRRI) rice varieties, application of recommended doses of fertilizers, and the adoption of modern technology. It led to the freedom from dependence on nature, which reduced the element of risk in agriculture. From the government side, the policy of prospecting in food grains was discouraged by the introduction of price support policies, at least for the food grains. The end result was that the index of agriculture production in Punjab increased from 230.0 in 1974/75 to 336.9 in 1978/79 (against the basic triennium ending 1961/62). At present Punjab contributes over 50 percent annually to the buffer stock of food grains, which is maintained by the government of India in varying quantities ranging from 15 million to 20 million tons.

The three major sets of factors responsible for rapid development have been extension education, the timely supply of inputs, and the price support policy. The Punjab Agricultural University established a great credibility among the farmers while making sound recommendations in agricultural technology. The university faculty, apart from addressing the farmers through mass media, made personal contacts with 50,000 farmers every year. The director general of the UN Food and Agriculture Organization recognized it as the largest extension training center in Asia.

The Punjab was known primarily as a wheat area, but in the last decade paddy cultivation has been taken up by the farmers in a big way. Rice procurement from this state is also the largest when compared with that of the other states of India.

This spectacular increase in total food production in Punjab in the last two decades is attributed to a number of technological variables. In this chapter crop-wise technology and its impact on the economics of various crop rotations will be assessed. The risk benefit, socioeconomic benefits, income disparities, and employment structure will also be examined.

## WHEAT

Wheat is the premier food-grain crop in the Punjab, and it occupies 67.3 percent of the net sown area. It is grown in the rabi season.[1] With the adoption of new technology associated with the introduction of dwarf Mexican varieties in the mid-1960s, the production of this crop has increased tremendously. Wheat production, which was 1.74 million tons in 1960/61, increased to 7.67 million tons in 1979/80. This fourfold increase in production in the last two decades was partly owing to the doubling of area under this crop but more owing to the increase in yield. The yield, which was 0.503 tons per acre in 1960/61, increased to 1.132 tons per acre in 1979/80. It coincided with the introduction of the dwarf Mexican wheats from 1965/66 to 1970/71 (see Table 9.2). The new varieties were resistent to lodging and were more fertilizer responsive when compared with the indigenous varieties grown earlier. Of the long list of new strains released during the last 15 years, some deserve special mention: PV-18 released in 1966, Kalyansona released in 1967, S-308 released in 1968, and WL-711 released in 1967 with potential yields of 1.80, 1.76, 1.70, and 2.12 tons per acre, respectively. The speed of adoption of these high-yielding varieties was very rapid, the area being 25,000 acres in 1967/68 and 70,000 acres in 1980/81 (as shown in Table 9.3). The recommended doses of 50 kilograms (kg) nitrogen, 25 kg phosphorous pentoxide, and 12 kg potassium oxide per acre are almost double those for traditional varieties. In 1965/66, very little fertilizer was applied to the indigenous varieties of this crop, but by 1978/79, when the new varieties had been widely adopted in Punjab, the average fertilizer used was 33 kg nitrogen and 19 kg phosphorous pentoxide per acre (see Table 9.4).

Before the introduction of new varieties, the extension agencies in the state had to canvass the farmers for the use of chemical fertilizers; now the adoption has reached a level that the fertilizer supply is not catching up with demand, in spite of a sharp rise in the fertilizer price.

It is well known that, under Punjab conditions, when high fertilizer doses are used, assured irrigation facilities are essential. This development in the state soon followed the widespread rural electrification. The number of tube wells and pumping sets, only 46,000 in 1965/66, increased to 565,000 in 1979/80. Whereas in 1960/61 only 58.2 percent of the cultivated area under wheat was ir-

TABLE 9.2

Area, Production, and Average Yield of Important Crops in Punjab, 1960/61 to 1979/80

| Year | Rice | | | Cotton | | | | Sugarcane | | | Wheat | | |
|---|---|---|---|---|---|---|---|---|---|---|---|---|---|
| | Area (in thousands of acres) | Production (in thousands of tons) | Yield (kilograms per acre) | Area (in thousands of acres) | Production (in thousands of tons) | Yield (kilograms per acre): Desi | Yield (kilograms per acre): American | Area (in thousands of acres) | Production (in thousands of tons) | Yield (kilograms per acre) | Area (in thousands of acres) | Production (in thousands of tons) | Yield (kilograms per acre) |
| 1960/61 | 567.5 (6.04) | 229 | 404 | 1,105.0 (11.89) | 120.30 | 109 | 109 | 329.0 (3.54) | 486 | 1,479 | 3,460.0 (37.26) | 1,742 | 503 |
| 1965/66 | 730.0 (7.67) | 292 | 400 | 1,134.0 (12.06) | 131.62 | 107 | 126 | 418.0 (4.44) | 570 | 1,365 | 3,830.0 (40.75) | 1,916 | 500 |
| 1970/71 | 975.0 (9.62) | 688 | 706 | 981.0 (12.90) | 147.31 | 137 | 161 | 316.0 (3.16) | 527 | 1,666 | 5,681.0 (56.72) | 5,145 | 906 |
| 1975/76 | 1,417.5 (13.64) | 1,447 | 1,021 | 1,433.0 (13.95) | 210.11 | 118 | — | 282.0 (2.74) | 613 | 2,167 | 6,027.0 (58.66) | 5,788 | 960 |
| 1979/80 | 2,930.0 (28.04) | 3,052 | 1,042 | 1,558.0 (15.07) | 204.87 | 97 | 163 | 190.0 (1.84) | 392 | 2,063 | 6,951.0 (67.30) | 7,869 | 1,132 |

Note: Figures in parentheses are the area under crop as percent of the total cultivated area.

Source: Punjab Government, Statistical Abstracts of Punjab (Chandigarh).

TABLE 9.3

Area under High-Yielding Varieties in Punjab, 1960/61 to 1980/81
(in thousands of acres)

| Year | Rice | Wheat | Cotton[a] | Sugarcane[b] |
|---|---|---|---|---|
| 1960/61 | n.a. | n.a. | 1,117 | 332 |
| 1967/68 | 10 | 25.0 | 1,049 | 342 |
| 1970/71 | 321 | 3,972.5 | 992 | 320 |
| 1976/77 | 1,485 | 5,940.0 | 1,387 | 282 |
| 1979/80 | 2,659 | 6,937.5 | 1,575 | 192 |
| 1980/81 | 2,718 | 7,000.0 | n.a. | n.a. |

n.a. = not available

[a]The medium and medium-long staple U.S. varieties (Gossypium hirsutum).

[b]The thick cane and tall cane varieties developed in the 1920s by crossing Sachharum spontaneum and Sachharum officinarum at Coimbatore (South India).

Source: Punjab Government, Statistical Abstracts of Punjab (Chandigarh).

TABLE 9.4

Crop Application and Recommended Doses of Fertilizers in Punjab
(kilograms per acre)

| | 1977/78 | | 1978/79 | | Recommended Dose | |
|---|---|---|---|---|---|---|
| | Nitrogen | Phosphorous Pentoxide | Nitrogen | Phosphorous Pentoxide | Nitrogen | Phosphorous Pentoxide |
| Rice | 35 | 6 | 40 | 8 | 50 | 12 |
| Sugarcane | 25 | 1 | 31 | 1 | 60 | — |
| Cotton | 15 | 2 | 20 | 2 | 40 | 12 |
| Wheat | 28 | 12 | 33 | 19 | 50 | 25 |

Source: Fertilizer Statistics, 1977-78 and 1978-79 (New Dehli: Fertilizer Association of India).

TABLE 9.5

Crops and Total Area Irrigated in Punjab, 1960/61 to 1979/80
(in thousands of acres)

| Year | Rice | Wheat | Sugarcane | Cotton | Total Area Irri- gated | Percentage of Net Area Irrigated to Net Area Sowed |
|---|---|---|---|---|---|---|
| 1960/61 | 460 | 2,014 | 235 | 1,040 | 4,992 | 54 |
| | (81.05) | (58.21) | (71.42) | (94.18) | | |
| 1965/66 | 630 | 2,608 | 346 | 1,082 | 5,592 | 59 |
| | (86.40) | (68.08) | (82.95) | (95.42) | | |
| 1970/71 | 885 | 4,796 | 270 | 1,205 | 7,137 | 71 |
| | (91.87) | (84.46) | (85.39) | (97.60) | | |
| 1975/76 | 1,315 | 5,403 | 242 | — | 7,707 | 75 |
| | (93.82) | (89.64) | (85.96) | (98.68) | | |
| 1979/80 | 2,816 | 6,494 | 170 | 1,414 | 8,792 | 85 |
| | (97.23) | (93.42) | (89.09) | (98.25) | | |

Note: The figures in parentheses indicate irrigated percentage of the crop area.

Source: Punjab Government, Statistical Abstracts of Punjab (Chandigarh).

rigated, by 1979/80 it increased to 93.4 percent (see Table 9.5). It is worth noting that the increase in irrigation facilities was most pronounced after 1965/66, when the new wheat varieties were introduced.

Another significant change noted in Punjab was that the conventional system of hoeing wheat was replaced by the application of such weedicides as Avadex, Tribunil, Dosanex, TOK-E, and so on. In other words, the farmers became conscious of the benefits of agrochemicals in general.

The completion of a network of link roads in the countryside during the last two decades also coincided with the price control policy introduced by the government of India in the mid-1960s. The Food Corporation of India and other government agencies lifted the entire stocks of wheat that reached the markets, and the practice of profiteering by middlemen was stopped. These marketing and pricing facilities provided direct incentives for the farmers to sell practically all the wheat soon after the harvest, and the usual 10 percent loss in rural storage was also saved. The market arrivals, which stood at 0.597 million tons in 1966/67, increased to 2.722 million tons in 1970/71 and 4.192 million tons in 1979/80. The publicly financed food-grain storage capacity, practically nonexistent in 1965/66, increased to 8.70 million tons in 1979/80.

There is no doubt that with the adoption of new scientific technology for cultivation of the wheat crop in the Punjab, the entire socioeconomic structure has changed. A number of factors, beginning with the consolidation of holdings and proceeding right up to the application of chemical fertilizers, assured irrigation through energized tube wells and pumping sets, and assured price and marketing facilities, has infused new confidence among the farming community of the state. The craze for new wheat varieties, even by the farthest villages, is so great that at one time they were willing to buy new seed by paying its weight in gold.

## PADDY

Paddy is a kharif crop in Punjab; in terms of net sowed area it stands next only to wheat, which is a rabi crop. In the past few years the area under paddy has been on the increase, replacing the less profitable summer crops. The total acreage of this crop constituted 6.04 percent of the cultivated land in the state in 1960/61 and increased to 28.04 percent in the year 1979/80 (see Table 9.2).

Traditionally, paddy was grown in the river floodplains and the waterlogged areas, but with the availability of underground water from the energized tube wells and pumping sets, the area has increased at a rapid rate. Among other contributory factors, the single largest incentive for the farmers has been the flat rate of electricity charged according to the horsepower of the motor rather than the actual units consumed.

Unlike the other tropical areas in India, where paddy is grown practically all the year round, only one crop is raised in this state. It breaks the cycle of pests and diseases; moreover, the intermediary winter further helps in this respect. Consequently, the crop yields are the highest in the country, being 1.56 tons per acre, compared with the national average of 0.8 tons per acre. Encouraged by the yield and the assured price, the farmers are adopting this crop very fast. The area under this crop, which was merely 567,500 acres in 1960/61, increased to 2,930,000 acres in 1979/80. The total production of unhusked rice increased from 0.229 million tons in 1960/61 to 3.052 million tons in 1979/80.

A number of semidwarf varieties with their origin at IRRI[2] have been introduced, and those have a high-yield potential. The Taichung Native-1 released in 1965, IR-8 in 1968, Jaya in 1971, and PR-106 in 1976 have a yield potential of 1.6, 2.2, 2.52, and 2.64 tons per acre, respectively.

As indicated in Table 9.3, the area under high-yielding varieties of rice increased from 10,000 acres in 1967/68 to 2,718,000 acres in 1980/81. Since the water requirements of this crop are very high, rice production is directly related to the improvement of irrigation facilities in the state. In 1960/61, only 81 percent of the area under rice was irrigated; this increased to 97 percent in 1979/80 (see Table 9.5). Like the Mexican wheat varieties, the IRRI rice varieties are also responsive to high doses of chemical fertilizers. In 1960/61 the application of fertilizers was negligible, but in 1978/79 it was 40 kg nitrogen and 8 kg phosphorous pentoxide per acre (see Table 9.4).

Other new innovations that have been adopted in recent years include the following: an optimum transplanting time for different varieties has been worked out; a rice nursery as old as 60 days can be transplanted under delayed seasonal conditions, such as the failure of monsoons; monocot weeds can be controlled by use of the weedicide Machete; leafhoppers and other pests can be controlled by effective and comparatively inexpensive insecticides; a number of precautionary measures have been suggested against the spread of bacterial leaf blight of rice. The price-support policy and the timely lifting of rice from the markets have been additional incentives. The actual price of a paddy is fixed according to a scale of the percentage of moisture in the grain, which also gives a fair deal of monetary benefit to the farmers.

Paddy production has also given a boost to the establishment of rice-based industries. The number of rice shellers, which stood at 85 in 1960/61, stands at 500 in 1979/80. Since most of the rice is exported out of the state, and the people of the rest of India prefer parboiled rice, this processing facility has now been added to about 10 percent of the shellers, and there are plans for further expansion. The production of rice increased from 14.61 thousand tons in 1960/61 to 197.37 thousand tons in 1978/79. At the present level, rice-based industries give employment to 7,000 workers, in addition to the employment of regular and casual labor for the cultivation, transport, and marketing of rice.

## WHEAT-PADDY ROTATION

Wheat-paddy rotation, which was not considered scientifically sound, was adopted by the farmers on their own. Their preference for growing these crops was based on economic considerations and the risk-benefit ratio as they compared with other crop rotations.

Wheat-paddy rotation is now quite common in the state and is gaining in popularity. The total production of food grains from one acre comes to 2.7 tons, which compares well with world standards. Based on the recommended inputs and potential returns of various items of cost, the returns over variable cost were worked out at current prices. The per acre gross return from wheat came to Rs2,384.5, while the various items of variable cost amounted to Rs1,372.7, showing Rs1,011.7 as the net return on variable cost. The return was higher than that from other rabi crops like gram, barley, rapeseed, lentil, and so on (see Table 9.6). In the kharif season, paddy also occupies a dominant position in terms of the return over variable cost (Rs2,230.0), while the other kharif crops like maize, bajra, and groundnut give much lower returns. U.S. cotton, on the other hand, competes well with paddy, but its price fluctuates too much from year to year, and it involves a high risk. Moreover, cotton is grown in the southwestern districts of the state where the climate is dry; the brackish underground water is not fit for irrigation; paddy crops in general are not grown, except in certain waterlogged areas; and sufficient canal irrigation water is available.

While planning a cropping system, a farmer not only considers the comparative profitability of different crop rotations but also gives due consideration to the degree of risk involved in terms of yield and price. In the case of wheat and paddy crops, effective price support is given by the government (commonly equivalent to the market price). Therefore, the degree of price risk in this rotation is minimized. In other crops, including pulses, oilseeds, sugarcane, and cotton, effective price support is not given.

The data on the harvest price and the average yields obtained from 1960/61 to 1979/80 were analyzed for the risk involved in the growing of different crops. Even though the degree of yield risk in wheat and rice is quite high, the price risk is quite low. Thus, the overall risk involved in wheat-paddy rotation, when compared with other crop rotations, is low.

## COTTON

Cotton is the most important kharif crop of the southwestern region of the state, where it is grown under canal-irrigated conditions. Punjab is the biggest producer of short, medium, and superior medium-staple cotton in India. In the past, the area under this crop has fluctuated significantly. For example, in 1960/61 cotton covered 11.89 percent of the net sowed area, where in 1979/80 it covered 15.07 percent. On the average, the state has only 7.79 percent of the total area under cotton in India, but it contributes 18.82 percent to cotton production. Although there has been a steady increase in the yield of U.S. cotton (except in 1979/80), the yield of desi cotton has been fluctuating with the change of climate.

A number of technological developments have helped the production of cotton in Punjab. The new U.S. cotton (Gossypium hirsutum) varieties, J-34 released in 1966 and F-414 released in 1977, with a potential yield of 5.5 and 6.0 quintals per acre, respectively, have a made their impact. Since U.S. cotton is attacked by a number of serious pests at various stages of growth, insecticide-spraying schedules were developed in the mid-1960s that effectively controlled leafhopper, whitefly, and cotton bollworms.[3] The difference in yield between the unsprayed and the sprayed cotton has been demonstrated at 2.5 times. New insecticides have replaced the old ones, and the latest group of pyrethroids have a great potential for the control of bollworms. The variety F-414 is of a shorter duration, and it vacates the fields two weeks earlier than the previous varieties; hence, it is possible to sow wheat in time and make the cotton-wheat rotation a success.

Trading in cotton, which until recently was entirely handled by private agents who manipulated price fluctuations by almost 100 percent, is now being handled by the Cotton Corporation of India. The price has stabilized around Rs450 per quintal.

Against the recommended dose of 40 kg nitrogen and 12 kg phosphorous pentoxide per acre, the average use of fertilizer at present is 20 kg nitrogen and 2 kg phosphorous pentoxide per acre. Likewise the recommended schedule of six sprays is rarely adopted, and the farmers barely apply three sprays. These gaps in the recommended and the actual adopted technology show that there is still a great chance for increases in production. Some of the newer innovations that still have a good chance to further increase production include the close spacing for high plant population, the treatment of seed with succinic acid for better seedling establishment, and the development of effective ground-spraying techniques. At present 20 percent of the area is covered by aerial spraying, and since the cotton fields are scattered, there is a rather limited possibility of further increasing spraying by this method.

Under Punjab conditions, cotton has the tendency to bear excessive vegetative growth during the monsoon season; therefore, efforts are being made to evolve sympodial varieties having the desired characteristics.

With the increase of production, several cotton ginning and pressing factories have also come up in the state. At present there are seven ginning and pressing factories, which handle 0.71 million bales of the total production of 1.22 million bales. Of the total production of about 0.33 million tons of cottonseed, only 10 percent to 15 percent is being processed for oil extraction; the rest is exported out of the state. The oil-extraction capacity in the state is being enhanced, and already six new mills are under construction. The existing seven spinning and two composite mills in the state are utilizing 8 percent to 10 percent of the cotton produced. When the seven new mills under construction come into operation, it is expected that about one-third of the cotton produced in the state will be processed locally.

## COTTON-WHEAT ROTATION

The return over variable cost for cotton works out to Rs1,385.53 per acre. For the cotton-wheat rotation combined the annual returns come to Rs2,397.27 per acre, as compared with Rs3,242.67 for the wheat-paddy rotation (see Table 9.6). The cotton-wheat rotation could give a much higher return, provided the local price of cotton is not allowed to crash. However, these two rotations are predominant in two different regions of the state, and their economics are complementary rather than being competitive.

Cotton-wheat rotation in the southwestern districts is possible if early varieties of cotton are grown for the timely sowing of wheat. At present the canal water available for irrigation does not permit 200 percent intensity of cropping; therefore, water-use efficiency needs to be improved. Although there is scope for conjunctive use of canal and underground water to a certain extent, the optimum requirements for various crop situations are yet to be worked out. Since paddy is quite resistant to salts, its cultivation might be possible and even considered necessary in these areas instead of cotton, particularly where the water table is already high or is rising steadily. Likewise, in the wheat-paddy areas there is a possibility of some shift to cotton-wheat rotation, provided the price of cotton is stabilized.

TABLE 9.6

Enterprise Budgets for Wheat, Paddy, Cotton, and Sugarcane in Punjab, 1980/81
(rupees per acre)

| | Wheat (WL-711) | Paddy (Pr-106) | Cotton (LH-372) | Sugarcane (COJ-67) |
|---|---|---|---|---|
| Gross returns | 2,384.50 | 3,996.00 | 2,580.00 | 8,800.00 |
| Cost items | | | | |
| Seed and seed treatment | 76.20 | 16.72 | 32.82 | 769.50 |
| Manures and fertilizers | 504.51 | 362.06 | 323.65 | 256.00 |
| Insecticides and pesticides | 154.75 | 254.73 | 269.40 | 677.62 |
| Irrigation | 128.10 | 480.00 | 90.00 | 200.00 |
| Human labor | 386.40 | 570.00 | 435.00 | 2,240.00 |
| Threshing | 51.24 | — | — | — |
| Cost of tying material | 20.00 | — | — | — |
| Marketing charges | 8.35 | 26.00 | 6.00 | — |
| Interest on variable costs (at 13 percent) | 43.21 | 55.56 | 37.60 | 134.65 |
| Total variable costs | 1,372.76 | 1,765.07 | 1,194.47 | 4,277.77 |
| Return over variable costs | 1,011.74 | 2,230.93 | 1,385.53 | 4,522.23 |

Source: Enterprise budgets prepared by the Department of Economics and Sociology, Punjab Agricultural University, Ludhiana.

## SUGARCANE

Sugarcane also occupies an important place in Punjab agriculture. The area under this crop has been fluctuating in the last two decades; 329,000 acres were cultivated in 1960/61; 418,000 acres in 1965/66; and 190,000 acres in 1979/80. Thus, it covered 1.8 percent to 4.4 percent of the net sowed area. Whereas the cultivation of paddy picked up in the state with the development of irrigation facilities, sugarcane with its high water requirements did not keep pace with it, owing to some other constraints.

The water, labor, and energy requirements of this crop are quite high. It is a preferred crop within the 10-kilometer areas of the sugar factories, where approximately one-third of the land comes under this crop. According to the Sugar Factory Act of 1953, the farmers there are given preference in the purchase of sugarcane. The high cost of inputs, the unstable price of sugarcane, and the frequent slumps in sugar prices are the major causes of fluctuation in the net sowed area of this crop from year to year.

Traditionally, sugarcane is known to be a high-value crop, and for many centuries the Doab[4] area of the state was famous for producing brown sugar, which traders sold throughout northwestern India. In the twentieth century, particularly after World War I, the cultivation of this crop acquired different dimensions. The Coimbatore varieties of sugarcane were developed by crossing Saccharum officinarum with Saccharum spontaneum. These hybrids spread throughout the tropical areas of the world, and varieties suiting the subtropical climate of the Punjab were also developed. Some of the famous varieties of the past that were popular throughout the state were COL-9, COL-312, COL-313, COL-285, COJ-46, COJ-67, and COJ-975. Practically all of them were midseason or late varieties; hence, sugar factories that opened early in the season would give low recoveries. These varieties also became susceptible to pests and diseases; with time those were replaced by newer varieties like COJ-1148 and COJ-64. The real breakthrough in growing sugarcane as a commercial crop in the Punjab came with the introduction of the early variety COJ-64; it has a sugar content of 16 percent to 17 percent and will be adopted by the farmers for a recovery of 10 percent to 12 percent. With the introduction of this variety in 1975, the factories started operating in the first week of November. The main-season variety COJ-1148 made it possible for the factories to adopt a varietal cane-crushing schedule and to prolong the running period.

From the point of view of the factories, the farmers should grow in equal proportion the early, midseason, and late season varieties. Unfortunately, there is no good midseason variety, and the main variety (COJ-1148) is getting susceptible to red rot and virus diseases. Therefore, new varieties are badly needed. On this will depend the future of sugarcane in Punjab, which climatically is only marginally suited for this crop.

Owing to the high cost of chemical fertilizers, the farmers at present apply only 50 percent of the recommended dose of nitrogen, which is 60 kg per acre. The variable costs for sugarcane come to Rs900 per acre. Nonavailability of labor for sowing, hoeing, harvesting, and transporting is already becoming an important constraint for further expansion. The seasonal labor that comes from outside the state is available only at the time of sowing and harvesting wheat, paddy, and cotton. On top of it all, there is too much fluctuation in the price of the sugarcane. Unless the price is controlled and it becomes remunerative to grow this crop, there are bound to be fluctuations in the area sowed.

New developments in the production technology—the hot-air treatment of cane seed to control the virus diseases; the application of insecticides for the control of borers, bugs, and pyrilla; the improved management of the ratoon crop; and the autumn planting of sugarcane—will be adopted by the farmers for greater production only if a remunerative price line is maintained. By the use of known technology it is still possible to increase by 50 percent the present yield of 204 tons of cane per acre and the total production of 3.92 million tons in the state. In the meantime, the existing six sugar factories, which have a daily crushing capacity of 7,500 to 7,750 tons, will have to be supplemented with four more factories in the next five years. It is also rather important to plan for a good production strategy.

## SUGARCANE AS ROTATION

Sugarcane is a long duration crop. It is sowed in February and harvested from December to March the next year. The crop is generally kept as a ratoon for one year more. The return over variable costs were worked out at Rs4,522.23 per acre in the factory area when the price of sugarcane paid by the mills was Rs25 to Rs30 per quintal[5] in 1980/81. Outside the factory area, the returns are much less. When the price paid by the mills was Rs13.00 per quintal in 1978/79, the farmers had to burn the standing crop; the cost of harvesting and transporting it to the factory was too high. Naturally, the farmers did not grow sugarcane the next year, with the result that during 1979/80 the total production of sugar in the country fell from the normal 6.2 million tons to 3.9 million tons. The cost-benefit ratio and risk-benefit ratio for this crop are unfavorable for sustained production.

The comparison of sugarcane with the wheat-paddy rotation (see Table 9.6) clearly illustrates that the cost-benefit and the risk-benefit ratios, though lower for the latter, find favor with the vast majority of farmers in the Punjab; hence, this rotation is becoming more popular. It will continue to be so until the cost of cultivation for paddy, owing to its high water requirements, becomes uneconomical or it is realized that the energy requirements can be more profitably utilized by growing some other crop. In other words, parameters like the nutrient balance and the energy input-output ratio are considerations related to national planning, while the cost-benefit ratio in terms of monetary returns directly concerns the farmers.

In general, sugarcane has a high level of price risk and yield risk. Therefore, this crop, even though it gives a high return on an average, does not find a sustained favor with the farmers in areas outside the factory zone. Hence, there is a periodic shortage of sugar, even though unlimited production potential exists in the state.

## GENERAL DISCUSSION

The three crop rotations discussed in this chapter cover almost 70 percent of the cultivated area of Punjab with a cropping intensity of 150 percent. Other crops, like vegetables, potatoes, oilseeds, fodders, pulses, and fruits, cover the remaining 30 percent of the cultivated area.

The recent shifts in cropping patterns reveal some very interesting features of Punjab agriculture. In the last five years, the area under desi cotton has decreased from 5.0 percent to 4.0 percent; that under corn has decreased from 13.9 percent to 9.7 percent; and that under oilseeds has decreased from 7.6 percent to 4.7 percent of the net sowed area. Naturally, there have been corresponding decreases in the production of these crops. The area under paddy has, on the other hand, increased in the last five years from 13.64 percent to 28.04 percent of the net sowed area in the state.

The cost-benefit ratio is, no doubt, of immediate importance to the farmers when choosing particular crop combinations, but it is not an ideal parameter because of the wide fluctuations in prices. Energy consideration offers a better alternative for making long-term comparisons. Data in Table 9.7 indicate that the energy output-input ratio for sugarcane is lower than that for wheat and paddy. Improved agricultural technology, meaning the increased use of fertilizers, improved seeds, better management practices, and mechanization, during the last two decades has brought improved returns to energy consumption in the production of sugarcane; it has not made much difference in the case of paddy. In the case of wheat the energy output-input ratio has declined from 24.4 in 1960 to 20.3 in 1979. This is due to increased energy input from machinery and agricultural operations, which has been matched by increased yields.

In Punjab, the increase in agricultural production was brought about by an improvement in irrigation facilities and the increased use of fertilizers, in addition to several other factors relating to credit and pricing policies. The improvement in irrigation facilities was brought about by a network of canals and by the sinking of tube

TABLE 9.7

Energy Balance for Wheat, Paddy, and Sugarcane

| | Paddy | | Sugarcane | | Wheat | |
|---|---|---|---|---|---|---|
| | 1960/61 | 1979/80 | 1960/61 | 1979/80 | 1960/61 | 1979/80 |
| Energy inputs (kilowatts per acre) | | | | | | |
| Machinery | 158 | 259 | 239 | 322 | 162 | 242 |
| Agricultural operations | 222 | 291 | 323 | 361 | 219 | 272 |
| Fertilizers | | | | | | |
| Nitrogen | 6.3 | 172.0 | 12.6 | 344.0 | 6.3 | 172.0 |
| Phosphorous pentoxide | 0.04 | 6.30 | — | — | 0.04 | 6.30 |
| Potassium oxide | 0.002 | 0.930 | — | — | 0.004 | 0.930 |
| Seed | 71 | 71 | 3,745 | 3,743 | 104 | 177 |
| Total | 458 | 792 | 4,319 | 4,772 | 434 | 871 |
| Kilocalories per acre | $0.39 \times 10^6$ | $0.68 \times 10^6$ | $3.71 \times 10^6$ | $4.08 \times 10^6$ | $0.37 \times 10^6$ | $0.74 \times 10^6$ |
| Energy output (kilocalories per acre) | | | | | | |
| Energy of grain | $1.63 \times 10^6$ | $4.21 \times 10^6$ | — | — | $2.01 \times 10^6$ | $4.52 \times 10^6$ |
| Fuel value of husk | $0.69 \times 10^6$ | $1.48 \times 10^6$ | — | — | — | — |
| Fuel value of straw | $5.76 \times 10$ | $8.44 \times 10$ | — | — | $7.08 \times 10^6$ | $10.80 \times 10^6$ |
| Energy of sugar | — | — | $4.85 \times 10^6$ | $10.16 \times 10^6$ | — | — |
| Fuel value of fiber | — | — | $9.03 \times 10^6$ | $14.23 \times 10^6$ | — | — |
| Fuel value of leaves | — | — | $12.32 \times 10^6$ | $17.20 \times 10^6$ | — | — |
| Total output | $8.08 \times 10^6$ | $14.14 \times 10^6$ | $28.64 \times 10^6$ | $39.98 \times 10^6$ | $9.09 \times 10^6$ | $15.17 \times 10^6$ |
| Output-input ratio | 20.6 | 20.8 | 7.71 | 9.74 | 24.49 | 20.35 |

Source: B. S. Pathak and Singh Daljit, "Energy Returns in Agriculture, with Specific Reference to Developing Countries," Energy 3 (1978): 119-26.

wells. Seepage losses from canals and the use of poor quality underground water lifted by the tube-well irrigation system have led to problems with saline and alkali soils. These soils are deficient in nitrogen, phosphorous, and zinc and contain a high concentration of boron, nobelium, and flourine.[6] The excess levels of flourine in these soils enter the food chain through plants, and the plant products damage human teeth. High levels of molybdenum in the saline and alkali soils, as well as in some waterlogged soils, result in the accumulation of this micronutrient in fodder crops. Cattle feeding on these fodders develop symptoms of molybdenosis disease, which is characterized by severe diarrhea and a loss of general health. Molybdenosis is actually a copper deficiency disease because excessive levels of molybdenum in cattle interfere with copper metabolism. Boron toxicity in plants is widespread in saline-alkali soils, but its effects on animal health are not known.

The excessive use of nitrogen fertilizers has caused the accumulation of high levels of nitrate in the underground water, a known health hazard. With the increased use of high-analysis fertilizers, unbalanced fertilization, and intensive cropping, deficiencies of zinc and sulfur are becoming quite acute for certain crops. In fact the deficiency of zinc is so widespread in paddy and wheat that zinc application has become an integral part of the crop production technology in Punjab.

With the expansion of irrigation facilities and the relatively better economic returns from paddy cultivation, even the sandy soils traditionally used for groundnut cultivation are being brought under paddy rather than growing irrigated groundnut.[7] Since the ponding of water in such soils is not easy and there is constant leeching, the paddy crop grown there exhibits typical symptoms of iron deficiency. These are some of the problems emanating from the use of modern agricultural technology, and in the future many more problems will, no doubt, come to the fore.

## SOCIAL DIMENSIONS

Along with the transformation of the agricultural economy in the past two decades, far-reaching changes are also occurring in the rural social structure of Punjab.

The per capita income per year of a typical farmer has increased from Rs902 in 1968/69 to Rs2,047 in 1977/78; hence, their standard of living has improved significantly. The farmers' increased income has been spent not on consumption items alone, but a major part has been plowed back into the farm and nonfarm productive assets (see Table 9.8).

TABLE 9.8

Change in Living Standards of Farmers

| | 1971/72 | 1977/78 |
|---|---|---|
| Investment of increased income by farmers on different items, 1968-78 (percent of total) | | |
| Farms assets | — | 50.57 |
| Nonfarm productive assets | — | 24.67 |
| Household assets | — | 14.67 |
| Expenditure on important food and nonfood items (percent share of income) | | |
| Food | — | 29.00 |
| Clothing | — | 11.41 |
| Education | — | 3.54 |
| Travel | — | 3.74 |
| Medicines | — | 6.40 |
| Liquor | — | 4.08 |
| Cosmetics | — | 1.00 |
| Ownership of certain conveniences (percent of farmers owned) | | |
| Television | — | 10.00 |
| Radio/transistor | — | 95.00 |
| Fan | — | 92.00 |
| Watches | — | 90.00 |
| Scooter/motorcycle | 32.0 | 57.00 |
| Children going to school (percent of total children) | 84.0 | 89.00 |

Source: P. C. Deb, D. R. Arora, and Singh Sukhminder, "Nature and Dimensions of Social Change in the Wake of Agricultural Breakthrough in Punjab," Ludhiana: Department of Economics and Sociology, Punjab Agricultural University, 1980. Mimeographed.

The use of modern technology in agriculture has not diminished the requirement for labor. Because of multiple cropping, it has increased most significantly during the peak seasons. To meet the increased demand of labor not available from the local labor force, the seasonal migration of labor from Uttar Pradesh and Bihar has become a common feature. The average wage of an agricultural laborer, which was Rs7.00 in 1967/68, has risen to Rs13.00 in 1980/81. Concurrently, some improvement is also noticed in the level of living of the landless agricultural laborers in the state.

Improvement in the standard of living is indicated by the construction of pukka houses and the possession of household conveniences, like electric fans, electric heaters, pressure cookers, and even refrigerators. The old mud houses have generally been replaced in part by pukka houses. There has been a significant increase in the ownership of personal conveniences, such as watches, radios, transistors, televisions, and scooters. There is hardly a farm household without a radio or transistor and a cycle now. The increased use of cosmetics and better dress, even by the landless poor, is a common sight—a reflection of the increased proportion of income being spent on these items. Side by side, the expenditure on liquor is also increasing, because it is affordable and because its use is considered a form of recreation.

The important factor in assigning social status in rural areas in Punjab in the past was the caste identity. However, now other factors, such as economic (agricultural) achievement and occupational status, are also gaining importance in determining social prestige. The functional utility of work is more valued now than the caste of a person. As a corollary to these socioeconomic changes, a modification in the caste hierarchy has also occurred in the villages. There is more flexibility in intercaste relations now replacing the rigidity of the system observed a decade ago in this region or still being observed in other parts of the country. Other evidence that the caste system is losing its grip on rural society comes from the fact that people in this region have started moving in large numbers from their hereditary occupations to economically more sound occupations or sometimes to the manufacture of value-added products.

There is an increased emphasis on the education of children, evident from the increased number of school-going children (see Table 9.8). Consequently, more expenses have been diverted to education by the rural families.

In general it can be concluded that the benefits of the so-called Green Revolution in the Punjab have been most significant, affecting all strata of life. The demand for the amenities of daily life has increased, and industry has also benefited. Owing to its peculiar geographical situation, little heavy industry has been set up in the state; hence, there is a nagging worry that, of the Rs8 billion worth of gross sales from food grains, only a small fraction is reinvested in the state and even a lesser amount is utilized for the further development of agriculture. Another socioeconomic omission is the inadequate use of skilled human resources for other parts of the country, where Punjabi farmers, given the facilities, can bring about spectacular improvements in agriculture.

## NOTES

1. Rabi season is a winter crop season, and kharif season is a summer crop season.
2. International Rice Research Institute, Los Banos, Philippines.
3. A. S. Atwal and Kuljit Singh, "Double Your Cotton Yield by Spraying," Progressive Farming 4 (1968): 4-5.
4. The land between the Beas River and the Sutlej River.
5. Early variety COJ-64 fetches a higher price.
6. J. S. Kanwar and N. S. Randhawa, A Review of Micronutrients in Soils and Plants in India (New Dehli: Indian Council of Agricultural Research, 1974).
7. Owing to the uncertain price of groundnut.

# PART III

## TECHNOLOGY CHOICES

### Application and Consequences

# 10

# MANAGEMENT: THE FORGOTTEN FACTOR IN TECHNOLOGY CHOICE

MICHAEL A. AMSALEM

> When choosing the equipment for our plant, we had to keep in mind that workers in this country have a low level of skill; we therefore had to choose machines stripped of all automatic features.
>
> Why such automated equipment? Because we are dealing with workers with such a low skill level that we had to choose equipment that would enable us to rely on them as little as possible.

Both of the textile plants whose choice of technology was described in these words by their managers were in Indonesia, close enough so that they could be viewed as drawing from the same labor pool. Both were producing an output of very similar technical characteristics in roughly identical quantities.

According to neoclassical economics, both of these managers could not have been right at the same time. Given the cost and productivity of the factors of production available to these plants, one of the alternative technologies available would have resulted in the lowest possible cost of production and should have been the one chosen by any "rational" decision maker.

Is that to say that at least one, and eventually both, of these managers were irrational? The temptation is strong to label "irrational" any behavior that cannot be explained through a rational and seemingly perfect model of the world. An alternative attitude is to wonder if the model against which real-world decisions are evaluated includes all of the factors influencing the choices studied.

Studies of technology choice in developing countries have long been confronted with numerous decisions that the neoclassical model failed to explain. This evidence has been so overwhelming that during the last 15 years economists have had to abandon the simplistic irrationality explanation and have tried to identify other factors, not included in the neoclassical model, that might influence technology choice in a rational way. Market distortions, cost of information, risk considerations, and government intervention have alternatively been invoked and have often been shown to explain some of the decisions made. Most of these new variables are factors that more or less directly influence the cost of the factors of production to the firm or the utility value of a distribution of potential returns.

There are, however, a number of factors that can influence technology choice by other means than by changing the relative price of the factors of production. Since these considerations have to do with the profile of the managers making the choice, the management process that leads to the choice of a technology, and management objectives and strategies, they can be grouped under the heading of "Management Variables" in technology assessment and choice.

Although most of these management factors have often been mentioned and some have even been studied in much detail, little effort has been made to bring them together and identify the different ways in which management variables can influence technology assessment and choice. Such a review is the objective of the following pages.

In an in-depth study of the technology choices made by 16 textile plants and 12 pulp and paper plants located in Brazil, Columbia, Indonesia, and the Phillipines,[1] as well as through a review of the numerous studies on the subject, it was found that management variables can influence technology choice in four different ways.

First, technology choice is the result of a management decision process. As such, it will be influenced by the procedures established by the firm for identifying and evaluating what are considered to be alternative technologies and for choosing from among them.

Second, whatever the level of quantification reached in the decision process, a large degree of subjectivity cannot be avoided, and the decision remains one made by specific individuals. As such, technology choice will be influenced by the background, skills, and prefences of the managers involved.

Third, alternative technologies require different types and different amounts of management skills for their efficient operation. Therefore, depending upon the characteristics of the management available to the firm, a given technology might be better suited than its alternatives.

Finally, the use of a specific technology will give a manufacturing facility a number of characteristics, such as flexibility, adaptability to changes in product design, and uniformity of output. These characteristics are important variables in designing a production facility able to respond to the market it is to serve. Depending upon the nature of its market and the strategy it adopts, a firm might prefer a technology for the specific characteristics it will impart to its manufacturing process.

This chapter will attempt to see if these management factors in the choice of technology influence the choices made by different firms in a consistent and predictable way.

## TECHNOLOGY CHOICE AS A MANAGEMENT DECISION PROCESS

The neoclassical model of technology choice is based upon several assumptions, including the following:

1. All alternative technologies and their characteristics are known to the decision maker at no cost to him;
2. Alternative technologies are characterized by their use of only two factors of production, capital and labor;
3. Minimization of production cost is the one and only criterion used by the decision maker in making a choice.

An examination of the decision processes followed for technology choices in the textile and pulp and paper industries leads one to question these assumptions and, perhaps more importantly, points to differences between these two industries. In the textile industry, equipment embodying alternative technologies is of a very standardized nature and, as such, is available on an "off-the-shelf" basis. Textile equipment manufacturers offer a certain number of models with well-defined characteristics only slightly modified by available options. Since inputs and outputs are fairly standardized, the operating characteristics of these machines and their performance data are well established and easy to verify. In this industry it can, therefore, be considered that alternative technologies and their characteristics can be known by the decision maker at little or no cost.

In the pulp and paper industry, on the other hand, equipment is normally custom-made for each facility. Very few pulp and paper plants are built in any one year, and each of these only requires one or a very small number of each type of equipment. Furthermore, wide variations in the characteristics of the inputs and of the final product sought make it technically unfeasible to satisfy the needs of several potential customers with the same equipment designs. For these reasons, it is uneconomical to develop standardized designs and streamlined equipment production facilities. For each pulp and paper plant, new equipment designs will be developed and the feasibility, cost, and performance data of alternative technologies will only be known to the decision maker after the cost of design and testing has been incurred.

The cost and availability of information on alternative technologies in the textile and pulp and paper industries, far from being characteristic of only these two industries, seem to be representative of two types of industrial processes. Mechanical processes, defined as processes that change the characteristics of the raw material through mechanical action, have few inputs, and these inputs have relatively uniform characteristics. Technological, as well as economic, factors limit the production capacity of machines used in mechanical processes. Consequently, increases in plant size are achieved by increasing the number of machines rather than by using larger ones. Equipment for mechanical process industries can, therefore, be standardized and produced in long production runs.

Chemical processes generally have a larger number of inputs. Furthermore, variations in the characteristics of these inputs have more impact on the design of the processing facilities than in mechanical processes. They also tend to have large economies of scale,[2] making it preferable to design a piece of equipment capable of processing all of the plant's production rather than to use several pieces of equipment of a smaller capacity. Equipment for chemical process industries will, therefore, need to be custom designed and built on specifications for each production unit.

The decision maker in the chemical process industries faces a situation in which a substantial investment has to be made to determine the characteristics of each of the technological alternatives he wants to consider in his choice. He will normally try to reduce the number of alternatives to be investigated by eliminating the ones that appear least promising. The problem becomes, then, that it is precisely the study he is trying to avoid that would provide him with the information needed to make such a decision. Faced with this dilemma, the managers in charge of technology choice in chemical process industries have been found to use a certain number of "rules of thumb" or a prioris.

These rules of thumb, in turn, point to another weakness of the neoclassical model of technology choice—its emphasis on capital-labor trade-offs and its failure to consider trade-offs among other inputs. Again this weakness is most obvious when contrasting the decision process in mechanical and chemical industries. In mechanical industries, capital and labor are the two most important inputs, and, although power sometimes play a role in specific operations, the minimization of production costs generally takes the form of a choice between alternative technologies using varying amounts of capital and labor. In chemical process industries, on the other hand, the consumption of power, steam, and chemicals normally plays a much more important role in the determination of production costs than does the use of labor.

In such industries, the decision maker will therefore tend to view technology choice as a trade-off between capital, power usage, and chemical usage rather than as a trade-off between capital and labor. Labor usage would tend to be a consequence of the choice rather than an important variable in the choice. When forced to eliminate alternatives, this decision maker will, as a rule of thumb, tend to retain those that involve trade-offs between important cost elements, namely, capital, power, and chemicals, and eliminate those that involve uncertain trade-offs between labor and some other input.

Up to this point, it has been assumed that minimization of production cost was the only objective pursued in the choice of technology and that that choice was made by one rational decision maker. An examination of the decision process followed by firms puts such a hypothesis in doubt, even if risk considerations are integrated into the evaluation of expected production costs.

For production cost minimization to be the only criterion in technology choice, each technology choice decision would have to be made independently. It was found, however, that decision makers tried to reduce the number of equipment suppliers for a given plant even if that meant choosing equipment that was not cost minimizing in some production steps. Such a reduction in the number of suppliers was thought by managers to ensure better after-sales service, as well as to reduce spare parts inventories and maintenance costs because of the standardization of simple parts by each equipment manufacturer.

A number of voluntary and involuntary commitments[3] were also made during the process of gathering information on alternative technologies and their characteristics. These commitments are particularly frequent in chemical process industries, where the costs incurred by an equipment supplier to generate the information needed by the potential client are high. The equipment supplier will only undertake such expenses when reasonably assured that an order will follow. Such commitments later constrain the final choice. Furthermore, some of these commitments are often contradictory and are resolved through a bargaining process internal to the firm between the different individuals or groups that made them. The result of such bargaining might then be a compromise that would be difficult to explain by a set of logical factors; this destroys the assumption of a single rational decision maker.

## MANAGEMENT PROFILE AND TECHNOLOGY CHOICE

Further arguing against this assumption of economic rationality are several other aspects of the decision-making process that make the decision dependent upon the training and skills of the managers participating in it.

When drawing up a list of alternative technologies to be considered for a specific operation, it was found that companies do not engage in a systematic search for the existing alternatives. Rather, the list consists of alternatives known to the managers involved in the decision. This practice appears to introduce a systematic bias in the choices made.

Typically, the managers involved in technology choice have been trained in developed countries, or their training has been heavily influenced by developed countries. Through their training they have become familiar with the equipment manufactured in these countries. To update their knowledge and keep abreast of the offerings of equipment manufacturers, these managers turn to trade publications and equipment exhibitions. Both of these tend to originate or be located in developed countries and, therefore, give more coverage to the equipment manufacturers of these countries. By nature these information sources also concentrate on innovations. Since innovations are made by equipment manufacturers with large research and development budgets and personnel, and these are generally developed-country firms, such a concentration further reinforces the bias in information flows. Finally, since developed-country equipment manufacturers are the largest, they are also the ones that can afford the promotion and worldwide network of representatives and agents needed to keep specific products in the minds of managers.

Because of these factors it was found that the managers involved in technology choice were well aware of the offerings of developed countries' equipment manufacturers but were by and large ignorant of even the names and locations of developing countries' firms manufacturing equipment for their industry. In the technology choices studied, only firms headquartered in developing countries with an equipment manufacturing industry included machines manufactured in developing countries in their list of alternatives.

If developed countries' equipment manufacturers offered the same range of technological alternatives, such patterns of information flows would not result in a bias in technology choice. But such is not the case. A review of the equipment offered by these two types of firms in the textile and pulp and paper industries clearly shows that developing countries' equipment manufacturers produce more labor-intensive technologies. Either because licenses become available from developed countries' manufacturers or because such technologies fall into the public domain, equipment embodying these labor-intensive technologies starts being manufactured in developing countries. At the same time, developed-country manufacturers under competitive pressure (or, more often, because of decreasing interest in their home markets) stop manufacturing these models and concentrate their efforts on innovations. Since they respond to what they perceive to be the needs of their main clients (the developed countries' equipment purchasers), their innovations will tend to take place at the more capital-intensive, automated end of the technology spectrum.[4] As a consequence the more labor-intensive technological alternatives were not even considered in most of the technology choices studied.

While the source of training and information available to the decision makers appears to have an impact on technology choice, so does the form of this training. Several studies[5] already point to the fact that engineering-trained and business/economics-trained decision makers will exhibit different patterns of technology choice. Engineering excellence, technological innovation, and construction quality will be criteria, whether conscious or unconscious, in the engineer's choice. Investment and production costs will play a much more important role in the businessman's decision. Since this is a well-researched phenomenon, it only needs to be mentioned here. However, it has generally been assumed that in firms managed by "economic men," the technology choices reflect their decision criteria of cost minimization. An examination of the decision process followed casts doubt on such an assumption.

Although "economic men" made the choice of technology decision, or at least were involved in it, in most of the firms studied they always worked on the basis of a list of possible alternatives that was established by engineers. Their choice was, therefore, already constrained to what the engineers considered to be acceptable alternatives. Discussions with engineers involved in the compilation of such lists of alternatives did not reveal the use of any clear guidelines (either given to them or generated by them) on how to select the technologies and manufacturers to be included in the list. It was rather a vague criterion of what the engineers considered to be "reasonable," to quote a word often used, that presided upon this selection, thus leaving full play to their technical biases.

Finally, managing machines requires very different skills than managing men. Here lies a very neglected aspect of the impact of managers' skills and profiles on technology choice. Attaining high efficiencies in the use of labor-intensive technologies requires labor-management skills: the capacity to organize, motivate, and control large numbers of relatively low-skilled workers. Attaining high efficiencies in the use of capital-intensive technologies, on the other hand, requires asset-management skills. Line balancing and production scheduling become of paramount importance, while the management of workers takes second place, usually restricted to the interaction of the manager with a relatively small work force of a higher skill mix and, therefore, professional motivation.

Whether engineering-trained or business/economics-trained, it is not surprising that managers who feel more confident in the management of machines will tend to choose more capital-intensive technologies, while those who feel more at ease with the management of men will tend to choose labor-intensive technologies. The traditional on-the-job training of managers, whether technical or business managers, tended to stress labor-management skills. The traditional boss, often the owner of the plant or with family ties to the owner, was somebody who had been brought up in the plant and trained there, knew the workers, and through time, had developed his worker-management skills. Technological development, both engineering technology and business technology, has made necessary the formal training of engineers and business managers. This formal training, conducted outside of the plant, tends, as will be seen later, to emphasize quantitative skills and machine management skills. When hired by developing countries' modern enterprises, these managers have little labor-management skills or experience and tend to exhibit a bias toward the use of capital-intensive technologies.

## MANAGEMENT REQUIREMENTS OF ALTERNATIVE TECHNOLOGIES

The relative emphasis on worker-management and machine-management skills of alternative technologies was shown above to influence technology choice by the bias it creates in managers of different profiles. In addition to this indirect impact on technology choice, differences in the management requirements of alternative technologies also influence technology choice in a direct way. As in the neoclassical model of technology choice, a firm, when choosing a technology, will try to optimize the use of the resources at its disposal. In contrast with the neoclassical, two-factor model, however, firms consider management as a resource in its own right and not aggregated into labor. Furthermore, it is not a uniform resource and needs to be characterized by types of skills. When choosing a production technology, a firm will then go through the rational, resource-use-optimizing process of searching for a technology that matches, among other factors, the amount and skills of the management it has at its disposal.

At this point it is necessary to elaborate on the meaning of worker-management skills and machine-management skills and look at the reasons why there are such differences in the skill requirements of alternative technologies. Two reasons for such differences were given in the previous section: the larger absolute number of workers required by labor-intensive technologies and the lower average skill level of these workers. While, by definition, labor-intensive technologies require a larger absolute number of workers, the issue of whether capital-intensive or labor-intensive technologies require a higher average level of skill from their work force has been a very controversial one.

Some authors argue that one of the reasons for the perceived preference of developing countries' firms for capital-intensive technologies is that these technologies substitute machine skills for worker skills. According to this argument, in countries where skilled labor is scarce, capital-intensive technologies, although they generate less employment, allow for the use of less-skilled labor. The opposite argument can also be found, namely, that as the level of capital intensity of equipment increases, so does the level of skill of the work force required to operate and maintain these machines.

In this study of technology choice in the textile and pulp and paper industry, it was found that the first argument, the one of decreasing skill requirement by technologies of increasing capital intensity, was correct if one only looked at machine operators. The drop in the skill level required from production workers is most important when moving from the most labor-intensive alternative for each operation to the next level of capital intensity. This was the case, for example, when even the most primitive power-spinning frames or looms replaced hand spinning and weaving. The reduction in required machine operator skills then becomes very small, if at all significant, when more capital-intensive technologies are adopted. Maintenance workers' skills, on the other hand, present a clear situation of the increasing skill requirements of more capital-intensive technologies.

When aggregating machine operators and maintenance workers, it was found that the overall skill level required of a work force dropped sharply when moving away from the most labor-intensive technology, but then kept steadily rising with increasing levels of capital intensity. For all practical purposes, however, it can be considered that overall skill levels increase with the degree of capital intensity, since the most labor-intensive technology is impractical to use in large-scale industrial ventures and is only suited to cottage industry. It was found that once the cottage industry type of labor-intensive technologies was eliminated, the most labor-intensive technology was, in most cases, the one using the smallest absolute number of skilled workers at a given production level.

Such differences in the skill mix of the work force required by technologies of varying capital intensity necessitate the use of different management techniques, depending upon the technology chosen. Labor-intensive technologies, once again excluding cottage-type technologies, require a heavy management emphasis on measurement of worker performance, worker motivation, and performance-related incentives. For capital-intensive technologies, the emphasis shifts away from direct incentives because of the higher degree of professional motivation of skilled workers and toward job satisfaction and technical training.

In their choice of a level of capital intensity for their manufacturing facilities, firms will tend to opt for the technologies that require the types of management techniques they feel most confident in applying. This is another cause of the preference for capital-intensive technologies exhibited by modern manufacturing enterprises in developing countries. Since the development of modern management techniques generally takes place in highly industrialized countries, more attention has been given to the design of policies aimed at increasing the efficiency of skilled workers rather than at increasing that of unskilled workers. The same emphasis is to be found in management education because of both the preeminence of developed countries' institutions in curriculum and course-material development and the tendency to teach the latest development in management techniques.

Although the effect of such a consideration is difficult to separate from the effect of other factors in technology choice, evidence of this phenomenon is found in a number of paired investments. Such is the case of two Japanese textile firms that started facilities of roughly the same size and type of output in the same country; one began with a strong and active local partner, well experienced in plant management; the other began with a weak and silent local partner. The first firm was found to have selected significantly more labor-intensive technologies and also was the one to show more confidence in its capacity to manage workers.

Up to this point, the management requirements of alternative technologies have been discussed in terms of skill requirements and management technique requirements. Alternative technologies also differ in terms of the number of managers required for a given level of output. In a given country, the number of top-level and middle-level managers required by a firm tends to depend upon the type of product manufactured, the market served, and, to some degree, the volume of production. The number of first-level managers (supervisors), on the other hand, was found to be proportional to the number of workers rather than to the volume of production. At a given production level, labor-intensive technologies will require a larger number of first-level managers than more capital-intensive technologies. Efficient and reliable supervisors were repeatedly mentioned by the executives interviewed as being extremely difficult to find and train, often more so than top-level managers. Because of the importance of their direct interaction with workers, the use of expatriates in such positions was not considered to be a viable alternative. It is, therefore, not surprising that a potential reduction in the number of first-level managers needed is a strong incentive for a firm to choose more capital-intensive technologies than warranted by the relative price of labor and capital.

## TECHNOLOGY CHOICE AND OUTPUT MANAGEMENT

The choice of a production technology, besides being a major determinant of production costs, will impart to a production facility a number of characteristics, such as adaptability to changes in product design, flexibility in terms of product mix, uniformity of output, and overall quality level of production. These characteristics are management variables in the sense that when planning a production facility, managers will have to decide on the optimal level to be sought for each of these variables. Such decisions will be based on an analysis of the market the firm intends to serve, an assessment of the environment in which it is to operate, as well as the likely accuracy of its market projections. Designing a production facility that rates the highest possible on each of these characteristics is an unrealistic goal for two reasons. First, there are generally trade-offs to be made among the desired characteristics, some of which might even be mutually exclusive. Second, there is usually a cost associated with improving any one of these features.

In a few rare cases, alternative technologies were found that differed in terms of some of these characteristics while still having similar levels of factor intensity. Such cases aside, however, variations in these characteristics are generally associated with variations in the level of capital intensity. For example, more labor-intensive technologies give a production facility more flexibility in terms of product mix than more capital-intensive ones. Because of the limitations imposed by manual intervention, labor-intensive technologies are generally embodied in equipment of lower production capacity per unit so that a larger number of units is required in order to attain a given level of output. Plants built to supply relatively small markets have to produce a wide product line in order to reach minimum economies of scale. Since managers of such plants will emphasize flexibility in terms of product mix when designing their facilities, they will tend to prefer more labor-intensive technologies. Labor-intensive technologies also impart lower fixed costs to a production facility and will, therefore, be preferred when management expects wide fluctuations in demand or has a low confidence in its market forecasts.

Capital-intensive technologies are generally thought to produce a more uniform output. One Indian plant chose an extremely capital-intensive technology for its production of parachute cloth, since uniformity of cloth quality was a major requirement. The same plant was using a very labor-intensive technology for its production of cloth used for industrial backing purposes, since irregularities in cloth construction were of little importance to the end-user.

The features that technologies of varying levels of factor intensity impart to manufacturing processes can be used, as seen above, to adapt a production facility to the characteristics of its market. Technology choice then becomes a function not only of factor costs but also of demand characteristics. This link between demand characteristics and technology choice can be considered to be reactive, since market characteristics are givens for the decision maker, and he merely selects the production technologies that best accommodate them.

Technology can, however, become an active variable in the implementation of a firm's strategy. A technology will then be selected for its capacity to differentiate the firm's output from that of its competitors. A technology chosen on such grounds will probably result in higher production cost than could have been achieved otherwise. If successful, however, such a strategy of product differentiation will position the firm outside of the price competitive segment of the market, enabling it to command a premium for its products, thus reducing the importance of production cost minimization.

Firms whose declared strategy was to use quality to differentiate their products were found to have chosen significantly more capital-intensive technologies than their competitors. They explained such a choice by their perception that more capital-intensive technologies produced an output of higher and more uniform quality. On the other hand, a firm whose strategy was to specialize in small orders to be filled on short notice chose equipment incorporating extremely labor-intensive technologies. Such technologies, because of their low investment cost, allowed the firm to maintain enough excess capacity to accommodate variations in demand. The large number of units also gave the firm the important flexibility to manufacture many product types simultaneously.

This dimension of technology choice as a variable in trying to achieve strategic objectives also seems to have influenced technology choice in the case of public companies. Public enterprises were found, in this study as in several preceding ones,[6] to have chosen substantially more capital-intensive technologies than their private counterparts. This phenomenon can be explained by the fact that, in evaluating the performance of public enterprises, volume of production is a more important criterion than cost of production. In the eyes of the managers of these firms, capital-intensive technologies were less dependent upon the operators' skills and accuracy and required a smaller and therefore more manageable work force. It is interesting to note that, because of poor maintenance and frequent breakdown, these capital-intensive technologies often did not provide the stability of production for which they were selected.

## CONCLUSION AND POLICY IMPLICATIONS

The foregoing description of the different ways in which management processes and characteristics influence technology choice resulted in a reexamination of a number of assumptions of the neoclassical model. The procedures developed by firms to identify alternative technologies and the problems they encounter in the assessment of these alternatives pointed to the weakness of the assumption of perfect information. The personal biases of the managers involved in the choice and the bargaining process within the firm that resulted from their voluntary and involuntary commitments led to a questioning of the assumption of rationality. The practice of considering management, both in terms of absolute numbers and in terms of types of skills, as a resource whose use has to be optimized rendered the two-factor neoclassical model much too restrictive. Finally, recognition of the fact that alternative technologies differ in more than their relative use of the factor of production and that they impart specific characteristics to production processes weakened the concept of production functions made up of totally interchangeable technologies.

While most of these findings are not new, they have not previously been presented as a set of management variables influencing technology choice. Although they have been mentioned in a number of studies of technology choice, their importance has often been discounted. Perhaps because of this discounting and perhaps because these different aspects of the impact of management considerations on technology choice have not been brought together, their potential implications for policy formulation have not yet been fully realized.

Advice given to developing countries' governments regarding policies aimed at promoting the use of labor-intensive technologies is still mostly based on the neoclassical model of technology choice. Recommended policies consist of lowering the cost of labor to the firm, raising the cost of capital, and, in general, reducing factor price distortions in order to bring market prices closer to their economic values. It is then assumed that these changes in relative factor prices will result in a shift of the firms' technology choices toward more labor-intensive alternatives. This, however, requires these firms to know about the more labor-intensive alternatives, to have the managerial resources that these alternatives call for, and to pursue the objective of cost minimization.

These policies may very well result in the choice of more labor-intensive technologies on an aggregate basis. In fact, there is strong evidence that relative factor prices do influence the overall level of capital intensity in a country.[7] There is, however, another set of policies that, although seldom considered, could result in a similar shift to more labor intensity. These other policies might also prove less expensive to implement, at least in specific environments, from both the social and the economic points of view. Such policies would work directly on the management variables earlier identified in an effort to increase the ability and the propensity of firms to choose more labor-intensive technologies at a given level of factor prices.

Furthermore, these two sets of policies, aimed at changing the relative price of the factors of production and at modifying the management variables that influence technology choice, would have a reinforcing effect on each other. While the first set would make it financially attractive to adopt more labor-intensive technologies, the second set would remove the management barriers to the adoption of such technologies. The simultaneous implementation of these two sets of policies would, therefore, have a greater impact than would the implementation of any one of them separately.

To overcome the information barrier to the consideration and eventual adoption of the more labor-intensive technologies, governments can, for example, take on the responsibility of searching for the more labor-intensive alternatives and disseminating this information to potential users. The existing technology centers or newly created ones could be put in charge of this search-and-information dissemination with the specific task of investigating equipment production in other developing countries from which the more labor-intensive alternatives seem to come.

To reduce the bias shown by developing countries' managers toward technologies whose successful operation rests upon machine management skills rather than labor management skills, developing countries could shift the emphasis to the latter in their management education. An increased supply of managers skilled at directing large numbers of lower skilled workers will also make it more attractive for firms to adopt labor-intensive technologies.

Finally, to increase the pressure on firms to adopt the cost minimizing technologies, governments should strive to increase the level of competition between firms. This could be achieved through decreased protection from imports, as well as a lowering of the institutional barriers for entry into industries in which there is already some amount of domestic manufacturing.

Such policies are essential in order to enable and motivate private firms to respond to relatively low labor costs by adopting labor-intensive technologies, all the more so if governments attempt, at the same time, to further lower the relative cost of labor.

NOTES

1. For a description of the methodology of this study and an analysis of its results, see Michel A. Amsalem, Technology Choice in Developing Countries (Cambridge, Mass.: MIT Press, forthcoming).

2. For the determination of the capital cost of equipment of similar design but varying scale in the chemical industries, engineers use what has come to be known as the "0.6 rule of thumb." It can be expressed as

$$K = a \cdot Q^{0.6}$$

where K = capital cost, Q = size, and a is a coefficient dependent upon the characteristics of the equipment. Although this formula has many limitations, it points to a less than proportional increase of cost with capacity. For an evaluation of the "0.6 formula" limitations, see Bela Gold, "Evaluating Scale Economies: The Case of the Japanese Blast Furnaces," Journal of Industrial Economics, vol. 23 (September 1974).

3. For a description of a similar pattern of voluntary and involuntary commitments in the analysis of foreign investment opportunities, see Yair Aharone, The Foreign Investment Decision Process (Boston: Harvard University Graduate School of Business, 1966).

4. A description of the process leading to an emphasis on capital intensity in developing countries' innovations can be found in Frances Stewart, "Technology and Employment in LDC's," World Development, March 1974.

5. For the first such study, see Louis T. Wells, Jr., "Economic Man and Engineering Man: Choice of Technology in a Low Wage Country," in The Choice of Technology in Developing Countries, ed. Peter Timmer et al. (Cambridge, Mass.: Harvard Studies in International Affairs, Harvard University, 1975).

6. See, for example, Wells, "Economic Man and Engineering Man"; and David Williams, "Choice of Technology and National Planning: The Case of Tanzania" (Doctoral thesis, Harvard University Graduate School of Business, Boston, 1976).

7. See, for example, Amsalem, Technology Choice in Developing Countries.

# 11

# TELECOMMUNICATIONS IN THE DEVELOPMENT PROCESS

DONALD M. LAMBERTON

Few technologies exemplify more fully the central problems of technology assessment than telecommunications. Technology assessment must go beyond cost-effectiveness and cost-benefit analysis of specific investment projects in a given socioeconomic context to the exploration of a wide range of scenarios that comprehend not only radical interrelated changes in technology and institutions but notions of sequentiality and continuity in those processes of change.

## TECHNOLOGY ASSESSMENT AND TELECOMMUNICATIONS DEVELOPMENT

The successful launch and return of Columbia, the world's first reusable spacecraft, understandably focused attention on the short-run technical performance: the test of success was reduced in the final hours to launch and landing. An article in the Economist ("Making Space a Local Stop" 1981, p. 95) acknowledged the potential global importance of the event but set out some sobering estimates of expected use: one-third military, one-third telecommunications satellites, and one-third yet-to-be-identified industrial users and scientific experiments. The article concluded, "The excitement of this reusable spacecraft will be not in the shuttle vehicle itself but in its cargo." In this way the Economist was reiterating the point made several years earlier by Clippinger (1977, p. 303), who argued that "information technologies are nothing more than mechanical means of distributing and processing information. Their end value, therefore, does not reside in their engineering or performance specifications, but rather in how well they address human needs."

Assessment of how well modern telecommunications facilities address human needs is obviously a very complex matter. The changes now taking place in these mechanical means of distributing and processing information are spectacular, and the potential changes are even more so. These changes not only address human needs in the sense of meeting existing demands but also change the pattern of demands, eliminating some and generating some that are entirely new. Several important implications should be noted. First, "information is necessarily about fragments of the cosmos and of history" (Shackle 1974, p. 10). Inevitably, attempts at assessment move beyond well-defined, optimization contexts to imaginative exploration of futures that are vague and often utopian. Such exploration is frequently condemned by policy analysts because it lacks a precise quantitative basis. However, as Freeman (1979, pp. 14-15) has reminded us, while technology assessment is extremely difficult, the attempt to explore a very wide range of implications is a necessary part of any democratic decision making.

Second, there is a need to look carefully to the organizational aspects. In the tradition of neoclassical economics it was held that useful analysis could be undertaken in terms of given tastes, given technology, and a given system of organization. It was not denied that organization was important. What was asserted was simply that there was no need to go into the business of inventing different ways of organizing. This is now challenged as telecommunications and other technologies change rapidly, generating changes in tastes, and exposing inadequacies in the organizational arrangements for assessing the effects of these processes. Corollary to this is the need for continuing organizational change. As Nelson points out (1977, pp. 143-44), there is no such thing as a final solution. This amounts to taking an evolutionary view of development and allowing flexibility in terms of both economic structure and modes of organization.

Issues of equity cannot be excluded from technology assessment of telecommunications in the context of this evolutionary process. Such issues underlie much of the New International Economic (and Information) Order debate, and it is widely recognized that most, if not all, major structural changes tend to accentuate inequality (Lamberton 1978, pp. 33-41). While there has been too little research into the matter, it seems plausible that communication systems may have a special role in relation to equity aspects of socioeconomic change. Insofar as information flows precede economic changes, the best informed may gain the greater benefits. There are also important differentials in information-using capacity. It is by no means clear that universal access approaches to the provision of mechanical means of distributing and processing information and information stocks will reduce inequality. On the contrary, equal access by individuals or countries of unequal information processing capacity may well accentuate the existing degree of inequality.

## THE COMMUNICATIONS GAP/SUITABILITY GAP THESIS

Against the background of these general observations about technology assessment of telecommunications developments, the question of appropriate technology will now be introduced. This can be done conveniently in terms of Streeten's communications/suitability gaps thesis (1981). Streeten argues that the growing inequality in the international distribution of income has come about because no international redistributive machinery exists to spread the benefit of the economic progress that has been achieved through a succession of scientific, technological, and industrial revolutions, which interacted with the human, economic, political, and social conditions that created and used those revolutions.

Analysis of labor productivity differences suggests the following are important: first, the mix of high and low productivity sectors; second, the proportion of output produced by best-practice methods; third, the extent to which best practice falls short of world best practice; and fourth, differences owing to economies of scale. If the gap arises from inadequate communication and/or transfer of existing technologies, Streeten labels it a communication gap. If the gap arises because of the absence of appropriate technology, it is labeled a suitability gap. While it is impossible to measure the relative importance of these two kinds of gap, the productivity analysis indicates that the differences (other than those owing to economics of scale) can be brought about by a communications gap. Attention should then focus on the ways in which telecommunications development could affect the communications gap.

The need for such inquiry is strengthened if a limitation of Streeton's discussion is considered. His paper concentrates on scientific knowledge and its application. He is, however, fully aware that entrepreneurial attitudes and the institutional setting are equally important. Therefore, the extent to which new communication systems address human needs can be assessed only if their contribution to the stimulation of the growth of knowledge is taken into account.

A catalog of effects could be developed. While diffusion theory approaches tend to characterize these as unmixed benefits that tend to raise labor productivity, improve the informational efficiency of the economy, or bring benefits in terms of greater social cohesion, there are negative consequences. For example, depending upon organizational arrangements, profits may be interposed between improvements in labor productivity and satisfaction of human needs.

A more important negative consequence concerns the suitability gap. Streeten hypothesizes that closing the communications gap may widen the suitability gap. The communications gap can be closed by improving the telecommunications system through lowered charges for given services, extended availability of given services, improved quality of service, or introduction of new types of services. The communications gap can also be closed by actions that complement the telecommunications services or otherwise facilitate the transfer of scientific knowledge and technology: building stocks of information, training information specialists, developing a body of scientific and technological expertise, and modernizing financial services. The negative consequence arises because, as Streeten says, "the more and the better advanced technology is communicated, the less likely it is that a new and more appropriate indigenous technology will be developed. Better communication reduces both the opportunity and the will to invent an indigenous technology."

The trade-off between the effects of these two gaps does not lend itself to an easy optimization solution. Even so, these changes reflect underlying economic conditions. The creation of information structures and stocks of information is capital formation in the economist's sense. As Arrow (1974, 1979) has argued, the combination of uncertainty, indivisibility, and capital intensity involved may mean that the pursuit of efficiency leads to rigidity and an inability to respond to further changes. It is, therefore, a mistake to think of the communications and suitability gaps as able to be switched on or off at will. The time period needed may be very long indeed; the legacies of colonial administrations may indeed by relevant. Furthermore, the irreversability characteristic may dictate a need for a virtual change of system. Conversely, it is necessary to recognize that policy results cannot be expected to be achieved quickly.

## MODELS OF TELECOMMUNICATION DEVELOPMENT

The next step in this chapter is intended to illustrate some types or models of telecommunications development experienced in the Third World. Developing countries have about seven percent of the world's 420 million telephones, 75 percent of which are located in only eight countries. The usual way of providing telecommunications facilities is by investment subject to commercial return criteria.

### The Commercial Model—Type 1

Papua, New Guinea (PNG) provides an excellent illustration of the commercial model—type 1. Karunaratne (1981) shows that with World Bank aid PNG modernized and revamped the interurban trunk routes that connect the copper mine, plantations, and centers of government administration. Eight urban centers with populations of more than 12,500 were linked to a main switching center using modern technology. Only 0.6 percent of the indigenous population in PNG are telephone subscribers. The system is dominated by the links with PNG's major trade partners. For them international communication in PNG is efficient. Given the World Bank mission assessment of PNG's system as "first-class," one might well wonder what its contribution to indigenous development will prove to be (Plant 1979). Other countries have had similar experiences (Clippinger 1977; Hai 1979; Valerdi 1977; Walsham 1979; Wellenius 1977). Reviewing the Canadian experience, Prentis (1978, p. 188) posed the question, Is telecommunications an instrument of economic development in a casual sense? Her judgment was that "there is little evidence to support a conclusion that telecommunication service by itself is a sufficient condition to stimulate or lead economic development. There is, nevertheless, overwhelming evidence that it is a necessary condition." The PNG experience does not dispute her conclusion. It would appear, however, that telecommunications developments in PNG have reinforced the existing pattern of development, and this would seem to be a consequence of the dominance of commercial criteria.

## The Commercial Model—Type 2

The second model for domestic telecommunications for developing countries might well be called the RCA model ("RCA Proposes Three Domestic Satellite Networks" 1980, pp. 7-8). Addressing the Third World Telecommunications Forum, Edgar H. Griffiths, president of RCA, said, "There has been a failure thus far to plan our satellite configurations in support of the most pressing of human needs. In the absence of advanced telecommunications whole regions and peoples remain ignorant of the simplest things that could improve their lives because there is no structured way to communicate to them. To do this would require cooperation and planning on a global scale.

According to Griffiths, a satellite system designed for international use is not the best for domestic purposes. The International Telecommunication Satellite Corporation satellites are located above oceans and are structured primarily for point-to-point telephony. Griffiths described the difference in the following way:

> They require a broad beam covering up to nearly one half the earth, are limited to low transmission power and so must have powerful earth stations with antennas no smaller than ten meters. On the other hand, domestic systems require satellites located above continents. Their beams

> would be concentrated on each user country or region. High transmission power in the satellite makes it possible to reduce ground station costs and the ground antenna can range from three meters down to one-third of a meter. It is clearly advantageous for developing countries to deploy many small and inexpensive ground facilities.

Griffiths believed that three domestic satellite locations could serve the needs of all countries in the world.

> One would cover both Americas. A second could serve Europe and Africa and include the Middle East. The third could extend across the Asian continent and include the islands in the Pacific Ocean. Each would require two or more satellites depending on use and the number of channels required—at the most a dozen satellites would would be needed initially. The technology is already here; and three clusters could be in operation in the coming decade. A high degree of coordination will be required to make these domestic systems efficient and cost effective. They must be standardized in equipment and function in space and on the ground; controlled by all participating stations. Otherwise domestic services might spring up at random—often more costly than necessary, possibly even the wrong system for the "wrong purposes."

Griffiths went on to state that the International Telecommunications Union (ITU) was the one organization with the experience, expertise, and international range to organize a world-wide satellite service, but it was hampered by lack of funds. Only 3 to 4 percent of United Nations Redevelopment Program funds were allocated for technical assistance in telecommunications. Even if this budget were doubled, a precondition for a global domestic service would be better public comprehension. "Perhaps the greatest single obstacle," Griffiths explained, "is the lack of understanding among government leaders on what specifically telecommunications can accomplish."

## The PEACESAT Model

Pending unraveling of the complexities, is there an alternative model? It has been argued that there is such an alternative (Hudson 1980, pp. 1B, 8); the Pan Pacific Education and Communication Experiments by Satellite (PEACESAT) began in 1971. They have linked some 17 locations in the Pacific Islands and the Pacific Rim, primarily in terms of an audio-conference facility, by means of Applications Technology Satellite-1 (ATS-1) (Bystrom 1975; Plant 1980). Disregarding the satellite and launch costs (ATS-1 was an old NASA satellite used for weather experiments), PEACESAT demonstrated the use of small, low-cost earth stations. Also it was significant because the information flow was two-way rather than one-way as in broadcast programs. Its principal objective was to demonstrate the benefits of currently available technology for the needs of sparsely populated, less-industrialized areas (Plant 1980, p. 1), more especially in terms of health, education, and community service programs.

Eulogies dominate the PEACESAT literature (Bystrom 1975; Kingan 1980). Plant (1980) is the first serious attempt at evaluation of its performance. Blocks of PEACESAT time are assigned to institutional users, including the Pacific Trust territories, the University of the South Pacific, and the PEACESAT network, which includes the above stations and others scattered from Hawaii to New Zealand and PNG. Various subnetworks were arranged for special interest seminars, emergency consultations, disaster relief, epidemic outbreaks, and earthquakes.

Plant's study sought to determine the predominant users and the relationships between them with special reference to the use of the system by Pacific Island people. Two broad types of quantitative analysis were undertaken: an analysis of PEACESAT exchange data and a network flow analysis of the PEACESAT system. The major objectives of the exchange data analysis were to determine the predominant categories of exchange on the system and the predominant locations and participants, including their sex and ethnicity. The network flow analysis was concerned with measuring the flow of information between and among the terminals in an attempt to determine whether any distinct relationships existed between them, especially between metropolitan and Pacific Island terminals (metropolitan terminals being categorized as those in developed countries, primarily Honolulu, Wellington, University of California-Santa Cruz, and Vancouver terminals).

Overall, the position of Pacific Island terminals relative to metropolitan terminals could at best be said to be weak, despite Pacific Island terminals outnumbering metropolitan terminals two to one. The figure for the amount of time on air used seemed to be a crucial one, and this showed a four-to-one ratio of Caucasian dominance over Pacific Islanders. But what was more disturbing was the role seen to be played by Pacific Islanders in the decision-making and administrative areas of the system, specifically the initiation and chairing of exchanges and the amount of time used on air by PEACESAT staff. There the role of Pacific Islanders was little more than marginal.

The results of the network flow analysis indicated a largely passive role for Pacific Island terminals, questioning less on the system while receiving information from metropolitan locations. Furthermore, the marginality of Pacific Island terminals was corroborated by the insignificant networking between Pacific Island locations and the inferior networking from Pacific Island to metropolitan locations.

More specifically, the Honolulu and Wellington terminals were overwhelmingly important: they used 54 percent of all air time, initiated 67 percent of exchanges, chaired 75 percent of exchanges, provided 65 percent question activity, and accounted for as much as 80 percent of all messages. U.S. nationals occupied over 40 percent of all air time (U.S. males taking 30 percent). A two-to-one domination of males over females was reported overall, but the sex ratio was more equally balanced in the Pacific Islander population. Administrative exchanges absorbed more than 20 percent of total time; 90 percent of this time allocation was to Caucasians. "This reaffirms the marginal role of Pacific Islanders in the decision-making functions of PEACESAT [Trans-Pacific Education and Communications Experiments Using Satellites]" (Plant 1980, p. 6).

Such a time analysis obviously falls short of what an economist would like to have (that is, values of the information flows). Nevertheless, Plant has raised serious questions about the usefulness of the PEACESAT model. Commercial criteria have not been applied. The project did not have to bear the initial satellite and launch costs, although it did operate with a continuing fear that the aging satellite would cease to function. The system was initiated, designed, developed, and operated by people with enthusiasm, technical skill, and integrity. But it cannot be judged to have contributed to indigenous development.

The outcome shows some similarities to the experience of the Universal System for Information in Science and Technology (UNISIST), UNESCO's global Science and Technology (STI) system (Lamberton 1978, 1979, 1980b). Guided by the objective of universal access and subject to the influences of professionalization, the single worldwide system view of UN agencies, and an inappropriate model of the development process, that system created a potential capacity that mirrored the existing development pattern. The content of the information system was not related to the needs of indigenous development.

On a lesser, but by no means unimportant, scale there are similar effects in the field of scientific publication. Gordon (1979) examined the problems of scientists working in developing countries. Those problems have recently attracted considerable attention as has the low level of scientific output. Gordon presented an examination of the editorial evaluation of papers developing-country authors are able to produce in an attempt to illuminate their problems through an analysis of their output. Developing-country authors were shown to experience far higher rejection rates than authors in developed countries (57 percent versus 17 percent) and to be faulted on specific points that tended to indicate that their major problems were those of maintaining current awareness. Gordon suggested this indicated that the problems developing-country researchers experience in accessing scientific information are a major factor in inhibiting the quality and quantity of their own scientific information output. Gordon raised the question, Quality measured by whom? In the journals investigated, 7 percent of authors were based in developing countries, but no one referee was. "It could be argued, therefore, that the dominance of peer review processes by scientists from technologically advanced countries discriminates against the interests of authors in [developing countries]" (Gordon 1979, p. 341).

## The International Telecommunication Union

The International Telecommunication Union (ITU) is the recognized international specialized organization for the international regulation of telecommunications; the establishment of equipment, operational, and tariff (charging) standards and practices; the coordination of telecommunication information for the planning and operation of services, including through the Union's Permanent Organs Headquarters (even on a daily basis); the establishment, as necessary, of the global and/or regional agreements on which detailed planning of individual services by each country can proceed with confidence; and the specialized agency responsibility within the UN family for the infrastructure development leading to an effective realization of the telecommunication systems necessary for information and communication transfer within and between nations.

The ITU has voiced the need to review the traditional approach in terms of commercial profitability, because that policy has favored the cities and large urban areas (Butler 1980). "Figure-conscious" decision makers had looked to direct profit returns and paid too little attention to the role of telecommunications in the infrastructure for national economic development. Changing telecommunications technology and changing national objectives have led the ITU to formulate the following recommendations and suggestions (Butler 1979, p. 21).

1. More attention should be given to the value of telecommunications as one of the instruments of development.
2. Developing countries should take advantage of the wider choice of telecommunication technology now available. Frequently,

the range of choice is even wider than for developed countries, who are committed to previous investments.

3. Transfer and adaptation of technology require judicious choice from the various options. The particular options should be chosen and applied in accordance with each country's realities and needs. Application of certain technology changes can only be phased in over long transition periods to avoid economic penalties to existing users.

4. Some telecommunication technologies, such as Direct Satellite Broadcasting and Computer Communications, which offer tremendous potential benefits for mankind, raise important political, socioeconomic, and other questions with negative elements. International statutes, including those of the ITU, and national legislation can contribute to satisfactory solutions. On balance, technical innovation should be restricted as little as possible because of the economic and cost penalties.

5. There is greater need for investment in telecommunications in the rural areas as a contribution to development and for a review of the criteria for fund allocation and return on capital in order to accelerate telecommunication development in these areas. Furthermore, a better appreciation is required of the indirect benefits, as well as direct benefits, of such investment. Appropriate research of the socioeconomic benefits should be undertaken.

6. New technology, a mix of shared satellite relay facilities and low-cost radio systems, could offer flexibility and relatively low-cost solutions in rural telecommunication problems, providing that there were adequate cost-sharing and operational partnership agreements between countries to service their rural areas by such innovative arrangements. They should at least be explored.

7. ITU has continued to keep abreast of technological development in establishing the necessary preconditions in the regulatory framework to permit the orderly adaptation or integration of new technology and technical methods into the World Telecommunication Network. All members should be encouraged to contribute further to these developments to service the needs of the rural areas.

These four models illustrate the diversity of experience and potential development. Not surprisingly, they reveal significant differences in the underlying perception of the environment and analyses of the processes of change along with conflicting interests. Translation of the complex technology into terms that the nontechnician can understand, as sought by the president of RCA, does not reach to the inequalities, dualism, and conflicts that are of great concern within the developing countries; it does not ensure achievement of the ITU's Integrated Rural Development objective; it does not alter the inequalities Plant reported in the PEACESAT case. It is of interest that the public good contribution of telecommunications investment was to have been explored in a major cooperative research venture launched by the ITU and the Organization for Economic Cooperation and Development (OECD) (Pierce and Jequier 1977). Little progress has been reported, perhaps more because of the complexity of the subject than the lack of support.

## ASSESSMENT OF TELECOMMUNICATIONS DEVELOPMENT

The ITU-OECD research project provided a blueprint for a comprehensive approach to the assessment of telecommunications development, albeit a somewhat biased one in that the balance of costs and benefits was judged to be favorable when the project was launched. While it remains largely an expression of intent, it can be contrasted with the much more restricted assessments sought by the World Bank. Saunders (1981) pointed out that the World Bank has sought to limit its role in the telecommunications sector to that of a lender of last resort. A further limitation arose from the attempt to focus on projects that would have impact on the lowest 40 percent income group. This appeared to divert funds from telecommunications to water supply, nutrition, and so on. This raises an important question, Is the achievement of such Rawlsian justice ensured without complementary development of communication services?

The starting point for assessment is recognition that communication activity is allocation dependent. To focus attention on benefits and ignore the costs can be quite misleading. Any discussion of satellite programs, for example, can be relied upon to produce statements that "we must have the best," that "we are moving toward the ultimate in satellite technology," and that satellite access is becoming "expensive." Such statements reveal a lack of awareness of resource limitations and the problem of social choice and a lack of historical perspective. More specifically, such comments reflect the fact that costs are to be met partially by someone else (for example, NASA carried the basic satellite cost of PEACESAT operations), a preoccupation with a dominant element of costs (for example, the spectacular downward trend of investment cost per circuit year of capacity from Early Bird to later Intelsat models), or simply a complete failure to reckon at all with costs.

The combination of economies of scale and cost reductions with new technology has generated expectations of change and the related belief that satellite service has become "inexpensive" and will become even more so. Such conclusions should not be based on a partial and misleading view. It is the cost of a unit of information flow that must be considered. Some relevant questions are the following:

Is it a flow to a "footprint" in a remote and sparsely populated area or to a coverage of one-third of the globe's surface?

Is it one-way or two-way?

Is it telephone or television?

Is it an information product the community wants or an alien one that may be rejected?

Can the recipients afford the ground facilities?

Can the recipients afford the associated software?

Can the recipients produce their own programs?

What economies of scale and cost reduction are apparent for software?

Can the size of market for the services (or for the software) be increased?

What time period is appropriate for cost calculations? Continuity can be vitally important for costs and for welfare (for example, in medical services).

Who meets the costs? Cross-subsidization is a common feature of communication services.

What alternative systems (including mixed systems) are available? Communication technologies are increasingly interdependent. Satellite costs may be going down, but so too may be the costs of alternative systems.

What external effects will the planned developments in communication technology have? Cost trends, such as that of investment cost per circuit year, look real winners, provided not too much attention is given to the history of the automobile.

What prior political choices are embodied in the cost allocation and pricing procedures of telecommunications authorities?

What is the role of competition in the provision of telecommunications services?

Is there a trade-off between communication and transportation of people and goods?

Is there a regulatory mechanism in operation? Does it need revision?

Who should make the decisions about the adoption of new component technologies?

Who should assess the demand for new services (for example, data transmission)?

Who should make allowance for equity effects?

Can new organizational arrangements be invented to overcome these problems?

Can those new organizational arrangements somehow be continually updated to avoid rigidity and inappropriate policies and to ensure "current" solutions?

To what extent do costs and benefits depend upon international coordination?

Technology assessment is not simply a matter of deciding upon an appropriate discount rate and using compound interest tables. The real difficulty is identifying the effects that bring costs and benefits, understanding the sequences in which they occur, and adding a time dimension to the scenario. Schumpeter (1953) thought in such a fashion. Society's "stock of goods," he said, "is neither homogeneous nor an amorphous heap. Its various parts complement each other in a way that we readily understand as soon as we hear of building, equipment, raw materials and consumer's goods. Some of these parts must be available before we can operate others; and various sequences or lags between economic actions impose themselves and further restrict our choices; and they do this in ways that differ greatly according to the composition of the stocks we have to work on" (p. 631).

Brief reflection suggests that information technology is treated as if it does not differ from manufacturing technology. Appropriate questions might be, How do the various kinds of information, both stocks and flows, complement both each other and other resources? Are these complementary relationships readily understood as soon as books, data bases, computers, land lines, and satellites are mentioned? Which parts must be available before others can be operated? What are the sequences and lags imposed by modern information technology? The answers are largely negative. This should not occasion surprise, as there is an element of truth in the accusation that economists have sought to analyze manufacturing using a production function concept derived from nineteenth century agriculture.

An enormous research task lies ahead for those wishing to undertake technology assessment of telecommunications development in a responsible way that will permit a rational social choice consistent with the degree of flexibility that is needed. Some indication of the extensive nature of such a research program is given in Hudson (1981), Wellenius' appendix to Saunders (1981), and Jussawalla and Lamberton (1981). Some topics central to such a research program are the following:

Information stocks and flows
Patterns of information use
The investment nature of information structures
The impact of new information technologies
Interregional and international information flows
Information labor markets
Cost analysis of information activities
Pricing of information goods and services
Information flows within organizations
Information activities and externalities
Policy-making processes for information activities

While this list emphasizes the economic aspects, the problems are properly addressed by interdisciplinary methods. Accordingly, the contributions of information science, political science, and sociology are needed.

There are two broad reasons why such research is not yet far advanced. The first is that there is a serious lack of statistical data relating to both domestic and international information activities. This, in turn, has restricted efforts to model the role of information in socioeconomic systems and to achieve a coherent picture of the information economy. At least the statistician can argue that many of the activities are as yet too new to warrant the collection of statistics. The policy maker has no such excuse and, instead of taking refuge behind the answer that there is no quick payoff, should be willing to explore the possibility that information-intensive socioeconomic systems may well behave in new ways and call for new policy approaches. There is general evidence that industrialized centers have become information intensive (OECD in press) and at least reason to suspect that developing countries are displaying a similar structural change; for example, service sectors in many developing countries are large and their investment in telecommunications is at a high level.

There is a certain novelty in treating the economy as an information system. Thinking members of the economics profession have long recognized that the price system was essentially an information mechanism (Lamberton 1980a). However, the information approach must be taken further if a proper understanding of the role of telecommunications is to be achieved. The Harvard University Information Resources Policy Program has pioneered new approaches that throw considerable light on the "information business" in an industrial society (Oettinger 1980). It is a matter of urgency that similar mapping of the pattern of change be carried out for developing countries. The U.S. experience suggests that so much of traditional modes of thought and organizational arrangements have now been challenged in the course of telecommunications evolution that it would be unwise not to expect similar pressures for change to emerge throughout the rest of the world.

This chapter has argued the need to know a great deal more than is presently known about the role of communication in development if technology assessment is to be applied to telecommunications development. This is now happening as part of the emergence of an economics of information (Lamberton 1971, 1974, 1978; Nelson 1980).

While evidence and opinion favors the judgment that information technology has the potential for change and the initiation of indigenous development, those who believe that this will be achieved are probably failing to understand the extent to which information resources can be controlled and can yield a differential advantage. As Rita Cruise

O'Brien comments, "At both national and intergovernmental levels, an enlargement of information capacity in the context of an organization for the South may alter this situation. But serious consideration is required of where and how electronic resources can be deployed most effectively. The present enthusiasm for informatics claims too much, both in terms of national planning and intergovernmental co-operation. Information systems themselves will not promote growth and development, nor will they strengthen the South in bargaining terms. Skepticism and critical reflection may direct resources where they will be best served and forestall the creation of new forms of disadvantage" (1980, pp. 47-48). Only intelligent use of "best practice" technology assessment techniques constitutes a defense against the pressures that can fritter away the potential gains and even accentuate the conflicts that characterize the development process.

## REFERENCES

Arrow, K. J. 1979. "The Economics of Information." In The Computer Age, edited by M. L. Dertouzos and J. Moses. Cambridge, Mass.: MIT Press.

_____. 1974. The Limits of Organization. New York: Norton.

Butler, R. E. 1980. Intergovernmental Conference for Cooperation Activities, Needs and Programmes for Communication Development. Paris: International Telecommunications Union.

_____. 1979. "The Application of Telecommunication Technology to Development." International Colloquium on Science, Technology and Society: Needs, Challenges and Limitations, Vienna, ITU, August.

Bystrom, J. W. 1975. "The Application of Satellites to International Interactive Service Support Communication." Proceedings of the Royal Society (London), pp. 63-80.

Clippinger, J. H. 1977. "Can Communications Development Benefit the Third World?" Telecommunications Policy 1 (September): 298-304.

Freeman, C. 1979. "Social and Economic Impact of Microelectronics." In Proceedings of the 1979 Workshop on Technology Assessment, edited by Department of Science and the Environment. Canberra: Australian Government Publishing Service.

Gordon, M. D. 1979. "Deficiencies of Scientific Information Access in Less Developed Countries." Journal of the American Society for Information Science 30 (November): 340-42.

Hai, S. 1979. "Telecommunications Development in Israel." Telecommunications Policy 3 (June): 134-46.

Hudson, H. E. 1981. "Toward a Model for Predicting Development Benefits from Telecommunication Investment." In Communication and Development: An Economics of Information Perspective, edited by M. Jussawalla and D. M. Lamberton. New York: Pergamon Press.

_____. 1980. "Satellite Communication in the Pacific Islands: Status and Strategies." In Proceedings of Pacific Telecommunications Conference, edited by D. J. Wedemeyer, Honolulu, pp. 1B, 1-9.

Jussawalla, M., and D. M. Lamberton, eds. 1981. Communication and Development: An Economics of Information Perspective. New York: Pergamon Press.

Karunaratne, N. D. 1981. "Telecom and Information in Development Planning Strategy." In Communication and Development: An Economics of Information Perspective, edited by M. Jussawalla and D. M. Lamberton. New York: Pergamon Press.

Kingan, S. G. 1980. "Appropriate Communications for Development." In Proceedings of Pacific Telecommunications Conference, edited by D. J. Wedemeyer, Honolulu, pp. 1-8, 21.

Lamberton, D. M. 1980a. "Communication as a Discipline: An Economist's Comments." Proceedings of Section 33, ANZAAS Congress (Adelaide), May 12-16, pp. 47-61.

_____. 1980b. "Economic Analysis of the UNISIST Experience." In Proceedings of First International Conference on Technology for Development, pp. 64-72. Canberra: Institution of Engineers.

_____. 1979. "Technology Transfer." In Directions for the Future: Communication, edited by Mari Davis, pp. 21-23.

_____. 1978. "Economics of Communication." In Planning Methods, Models and Organization, edited by S. A. Rahim, D. M. Lamberton, et al., pp. 21-97. Honolulu: East-West Center.

Lamberton, D. M., ed. 1974. The Information Revolution, vol. 412. Philadelphia: Annals of American Academy of Political and Social Science.

_____. 1971. The Economics of Information and Knowledge. Harmondsworth: Penguin Books.

"Making Space a Local Stop." 1981. Economist, April 4, p. 95.

Nelson, R. R. 1980. "Production Sets, Technological Knowledge, and R&D: Fragile and Overworked Constructs for Analysis of Productivity Growth." American Economic Review 70 (May): 62-67.

_____. 1977. The Moon and the Ghetto. New York: Norton.

O'Brien, R. C. 1980. "Specialized Information and Interdependence." Telecommunications Policy 4 (March): 42-48.

Organization for Economic Cooperation and Development. In press. Economic Analysis of Information Activities. Paris: OECD.

Oettinger, A. G. 1980. "Information Resources: Knowledge and Power in the 21st Century." Science 209 (July): 191-98.

Pierce, W. B., Jr., and N. Jequier. 1977. "The Contribution of Telecommunications to Economic Development." Telecommunication Journal 44: 532-34.

Plant, C. M. 1980. "Telecommunications in the Pacific Islands: Helping or Hindering Development?" In Proceedings of Pacific Telecommunications Conference, edited by D. J. Wedemeyer and D. L. Jones, pp. 1D, 12-18. Honolulu.

Prentis, M. R. 1978. "Some Economic Considerations in Planning Telecommunications Services Development in Isolated and Underprivileged Areas of Canada." Proceedings of Workshop on Special Aspects of Telecommunications Development in Isolated and Underprivileged Areas of Countries, pp. 179-88. Ottawa: Department of Communications.

"RCA Proposes Three Domestic Satellite Networks to Serve Worldwide Needs." 1980. Information Hotline 12 (January): 7-8.

Saunders, R. J. 1981. "Telecommunications in Developing Countries: Constraints on Development." In Communication and

Development: An Economics of Information Perspective, edited by M. Jussawalla and D. M. Lamberton. New York: Pergamon Press.

Schumpeter, J. A. 1953. History of Economics Analysis. New York: Oxford University Press.

Shackle, G. L. S. 1974. "Decision: The Human Predicament." In The Information Revolution, edited by D. M. Lamberton, vol. 412, pp. 1-10. Philadelphia: Annals of the American Academy of Political and Social Science.

Streeten, P. P. 1981. "The Conflict between Communication Gaps and Suitability Gaps." In Communication and Development: An Economics of Information Perspective, edited by M. Jussawalla and D. M. Lamberton. New York: Pergamon Press.

Valerdi, J. 1977. "A Communication Plan for Mexico." Telecommunications Policy 1 (September): 271-88.

Walsham, G. 1979. "Models for Telecommunications Strategy in the LDCs." Telecommunications Policy 3 (June): 105-15.

Wellenius, B. 1977. "Telecommunications in Developing Countries." Telecommunications Policy 1 (September): 289-97.

# 12

# THE ECONOMIC FEASIBILITY OF EMPLOYING USED INDUSTRIAL MACHINERY IN DEVELOPING COUNTRIES

DILMUS D. JAMES

## INTRODUCTION

There is a thriving trade in secondhand industrial machinery in the more developed market economies, some of which (enough to make a significant contribution to industrial capital formation) finds its way to developing nations. The only estimate of the aggregate contribution of used machinery to industrial equipment formation for developing countries puts the proportion between 10 and 20 percent (James 1974, pp. 95-97). This rough estimate, based on an extrapolation of several categories of used machinery exported to developing countries from the United States, is subject to wide margins of error, but 10 percent would appear to be a conservative figure. Other observations would appear to corroborate this order of magnitude. W. Paul Strassman found in a sample survey of industrial firms that about one-half of Mexican and three-fifths of Puerto Rican firms employed used machinery for major processes (1968, p. 207). Constantine Vaitsos reported that a "considerable part" of the machinery imported by Columbia is secondhand (1970, p. 38). Maarten Pronk surveyed production sharing or maquiladora plants in Ciudad Juárez, Mexico, and discovered that, excluding the building, the percentage of total investment consisting of used machinery was 76 percent, 61 percent, and 34 percent for shoes and clothing, electronics, and other plants, respectively (1981, p. 46). Charles Cooper, Raphael Kapinsky, and R. Turner mentioned that since 1965 the Dundee, Scotland, market for used machinery has sold secondhand jute mills and weaving plants to African countries, Bangladesh, and Thailand (1975, p. 131). Personal correspondence with a number of observers in India, particularly Ishwar Dayal, who furnished data from a study by B. K. Hedge, revealed a heavy usage of secondhand equipment in India's cotton textile industry.

TABLE 12.1

U.S. Exports of Selected Industrial Equipment to Developing Countries
(in thousands of dollars)

| Year | Total New and Used Equipment | Used Equipment | Percent of Used to Total Equipment |
|---|---|---|---|
| 1970 | 24,496.6 | 7,464.1 | 30.5 |
| 1971 | 29,017.2 | 9,174.4 | 31.6 |
| 1972 | 33,401.1 | 11,017.6 | 33.0 |
| 1973 | 46,817.8 | 15,636.4 | 33.4 |
| 1974 | 68,726.3 | 23,342.3 | 34.0 |
| 1975 | 87,696.1 | 21,886.5 | 25.0 |
| 1976 | 109,021.1 | 23,448.5 | 21.5 |
| 1977 | 82,970.4 | 19,209.0 | 23.2 |
| 1978 | 89,352.0 | 21,576.0 | 24.1 |
| 1979 | 89,045.9 | 23,138.7 | 26.0 |
| 1980 | 117,993.8 | 24,111.4 | 20.4 |
| Total | 778,538.3 | 200,004.9 | 25.7 |

Note: "Equipment" includes metal-cutting machine tools, metalworking machine tools (changed to "metal-forming" in 1978), industrial sewing machines, and paper and pulp mills.

Source: U.S., Department of Commerce, Exports, Schedule B, Commodity by Country, 1970-1977, FT 410; 1978-1980, FT 446.

As can be seen from Table 12.1, a significant amount of industrial equipment is exported to developing countries. Used and new industrial machinery exports by the United States are distinguished for four categories: metal-cutting machine tools; metalworking machine tools (a designation changed to metal-forming in 1978); industrial sewing machines; and pulp and paper mills. For the years from 1970 to 1980, about one-fourth of these types of equipment exported to developing countries was secondhand. As far as the author knows, no other nation's official data regularly distinguishes new from used equipment for any important industrial capital goods.

Regarding the global used machinery market and the total amount traded, very little is known. Based on a limited sample survey,

S. W. Tomalin, director of the U.K.-based Association of European Machine Tool Merchants, reported an estimate of approximately $600 million worth of used machine tools sold in the 1978 U.S. market and $150 million worth for the U.K. market. Considering inflation and market growth, Tomalin guesses the 1979/80 fiscal year sales of used machine tools of around $1 billion for the United States and $200 million for the United Kingdom. It would appear, then, that a reasonable guess for 1980 would be a global market for used machine tools of around $2 billion. Trade sources suggest that approximately 5 percent of U.S. and 8 percent of U.K. secondhand machine tools are exported, with about one-half of each nation's exports going to developing countries. Assuming a U.S. market proportion to developing countries at 2.5 percent and using the U.K. 4.0 percent proportion for the rest of the world, we arrive at 3.25 percent of global used machine tool sales going to developing countries, or roughly $65 million.

Note that this estimate includes only used machine tools. Venturing beyond this point is extremely hazardous without further empirical evidence; a single secondhand plant can sell for millions of dollars.[1] The two nonmachine tool components of the four categories of U.S. industrial used machinery exported, used industrial sewing machines and used paper and pulp mills, averaged about one-fourth of the total used machinery exports reported. Despite all of the uncertainties, a figure in excess of $100 million for total used machinery transferred to developing countries would appear to be conservative indeed. Without more intensive gathering of data, this is probably the best guess that can be mustered. Even, however, allowing for wide margins of error, the data indicate that a serious investigation of the transfer of used machinery to developing countries is warranted.

The importation and employment of used machinery by developing countries has not been entirely ignored, although one can hardly refer to a copious literature. From the late 1950s to the mid-1960s several studies appeared, including those sponsored by the Netherlands Economic Institute (1958), the Rand Corporation (Moore 1961), the Economic Intelligence Unit (1962), the Agency for International Development (Ralph M. Parsons Company 1965), and the United Nations (1966). During the same period, Andrew Schonfield (1962) and Albert Waterson (1964) recommended that the feasibility of secondhand equipment for developing countries be given serious scrutiny, and A. K. Sen (1962) published what still constitutes the main theoretical elements for explaining the attractiveness of used machinery to Third World countries. Aside from the brief espousal of used machinery by Strassmann (1968, pp. 203-13), little was published on the subject in the late 1960s; international agencies became inactive when official response from developing countries to the UN report was unenthusiastic. Since the early 1970s there has been some revival with a case study of secondhand jute processing equipment in Kenya (Cooper, Kapinsky, and Turner 1973), cost comparisons of used and new construction equipment (Schwartz 1973), a case study of a secondhand automated paper plant in Mexico (James 1973), a book-length study of used machinery in developing countries (James 1974), and a highly theoretical treatment by M. A. M. Smith (1975). Howard Pack (1978) added a study of used machinery for the cotton textile industry.

This chapter will support the general conclusion distilled from the foregoing literature: a judicious use of secondhand equipment can benefit individual entrepreneurs in developing countries, as well as further overall economic development objectives. Furthermore, simple and relatively inexpensive alterations in the used machinery market and policies influencing secondhand equipment use in developing countries could lead to a significant expansion in used machinery transfers to Third World countries. Essentially, the present study represents a resume of an earlier book, Used Machinery and Economic Development, with updated material gathered in 1981 through correspondence and personal conversations with used machinery dealers' associations, individual dealers, and developing-country users. Field research was also conducted in Northern Mexico by Lucinda Vargas, a research assistant.

The chapter will proceed by reviewing the private motivations for employing secondhand equipment. Also a variety of social returns can be reaped from producing with used machinery, benefits that ordinarily will not be considered by an individual entrepreneur. These community benefits are also explored. The chapter further reveals a host of market imperfections and biases surrounding the transfer of used machinery to developing countries, the vast majority of which, unfortunately, tend to suppress the amount of secondhand equipment employed. Finally, conclusions and policy implications are presented, as are suggestions for further research.

## RATIONALE FOR EMPLOYING USED MACHINERY

What motivates business entrepreneurs in developing countries to consider opting for used machinery that has been sluffed off in the more industrial countries? The main theoretical justifications (Sen 1962; Smith 1975) rest on the following propositions: labor is relatively cheap and capital relatively expensive in developing countries vis-à-vis developed countries; earlier vintages of technology are likely to incur higher labor and lower capital costs vis-à-vis recent vintages, since the general thrust of technology tends to be labor saving and capital using; and after years of use, older vintages of capital equipment will ordinarily require the application of increasing amounts of repair and maintenance, activities generally considered to be relatively labor intensive. Under these conditions, the net value of a given piece of equipment will tend to fall more rapidly in the more industrial countries than in developing countries.

Three questions remain to be explored. First, given the strong theoretical case for employing used machinery in developing countries, why is new equipment imported at all? Second, after allowing for real world limitations inherent in using secondhand capital goods, can the fundamental theoretical underpinnings, based on differences in factor costs in developing countries and developed countries, explain the volume of used machinery actually transferred? A negative answer here brings up the third problem; namely, if the usually adduced theoretical considerations cannot fully account for the volume of imports of used machinery to developing countries, what additional ingredients should be incorporated into the analysis?

Using a number of restrictive theoretical assumptions, M. A. M. Smith (1975) arrived at the conclusion that developing countries would be better off importing all used machinery. Needless to say, as of course Smith recognizes, a variety of practical considerations intervene in the real world. In addition to a whole catalog of imperfections in the sequence of market transactions transferring used machinery to developing countries, a matter to be broached later, there are a variety of legitimate reasons why, on a case-by-case basis, new machinery will prevail. Older equipment can result in a loss of product quality or wastage of material inputs as appropriate operating tolerances deteriorate. Technological progress is often capital saving per unit of output, as well as labor saving; thus recent vintages of capital goods might be more appropriate for factor endowments in developing countries. Problems with spare parts availability and low levels of maintenance skills in Third World countries can predominate over other potential economies of secondhand equipment. Obtaining information on new machinery is usually less costly, and follow-up guarantees and servicing agreements are more routine and reliable. While not intended to be definitive, these examples illustrate the point that, although used machinery can be a bargain, it is not necessarily so.

Indeed, factor price differences alone probably cannot explain the extent of used machinery moving to developing countries. As an illustration, data for 1950 and 1960 vintages of cotton textile equipment in Latin America, compiled by the Economic Commission for Latin America (1966), were adjusted by a range of lower capital costs and a range of higher operating costs in order to obtain a matrix of hypothetical unit costs of production associated with employing used machinery in developing countries.[2] Ratios of used to new machinery costs assumed were 50 percent, 75 percent, and 90 percent. Capital costs represented not only the purchase price of the equipment, it also embraced any other necessary costs associated with getting the used machinery ready for operation (for example, inspection and appraisal, reconditioning, and so on). Increases in variable maintenance costs of 50 percent, 100 percent, 150 percent, and 200 percent were assumed for used machinery compared with new equipment, with the figures for variable maintenance cost standing as a proxy for any increased operating costs associated with employing the older equipment.

The resulting matrix of unit costs of production was then compared with 1965 vintage equipment, which from the vantage of the 1966 study, was the new or most recent vintage of technology. The results of this case study are instructive, and while no claim to universal applicability can be made, the hypothetical range of adjustments for used machinery seems reasonably representative of most industrial equipment. The 1950 used machinery was not competitive with 1965 new equipment for any assumed value. The 1960 used machinery was more efficient for the 50 percent cost of capital figure for all assumed rises in operating costs. The only other combination for which used machinery triumphed was with the lowest rise in operating costs and a 75 percent (25 percent reduction) in capital costs. Since the acquisition of five-year-old capital goods in the textile industry at 50 percent of original cost is unlikely, especially if indirect costs of inspection and appraisal and reconditioning are included, only one of the posited permutations appears to be both realistic and feasible.

But, if theoretical fundamentals in isolation can account for only a small transfer to developing countries, what accounts for approximately one-half of used machinery exports going to these nations? Over ten years of researching and observing the use of secondhand equipment in the developing countries has led to the conclusion that the answer lies in a variety of unusual or special situations, which, although not important individually, become so when viewed in the aggregate. In virtually all of the cases of used machinery employment in developing countries that have been learned about in any depth at all, one or some complementary combination of these "special situations" has been present to reinforce the basic differences in factor costs. Without attempting to cover the entire range of special situations, a few can be mentioned illustratively.

Ordinarily, used machinery will involve higher maintenance costs compared to vintages of technology owing to greater age and use coupled with the trend to build more durability into newer equipment. This is by no means always the case, however. As an example, a dairy owner from northern Mexico bought a used compressor in Lansing, Michigan, which had been rendered obsolete in the United States by more sophisticated equipment that could cool the milk more rapidly and increase throughput. Although he purchased the used compressor at "a dime on the dollar" and incurred only minor repair costs, the primary reasons for purchasing the secondhand item were that it was simpler, more rugged, had few moving parts, and was free of complicated electronic controls. The expectation of lower maintenance expenses was the entrepreneur's express reason for scanning the available equipment in the U. S. used machinery market.

Other special situations abound. While it is usually true that new machinery will last considerably longer than used, extraordinary durability of older equipment can eliminate this as a consideration. The author once observed in an Oneida silverware factory in Toluca, Mexico, the operation of an enormous 20-ton drop hammer, approximately 50 years old, which was mechanically raised and fell by gravity. It is virtually impervious to physical wear and tear. In addition, smaller markets in developing countries will sometimes be a factor for opting for used machinery, since earlier vintage technology implies a lower optimum volume of output. When equipment has a low utilization factor owing to seasonal demand, unpredictable peak demand periods, standby equipment for emergency use, standby equipment to temporarily replace a machine being overhauled, or similar reasons, the higher capital costs of new machinery can tip the balance in favor of employing used machinery for these purposes. Similarly, a one-time production task that is relatively short-term in nature can be more advantageous if used machinery is selected. Modern Plastics ("Why Used Equipment Is Selling Again" 1977) cites short runs as one reason for buying used plastic-producing equipment. Fuel changes can be a factor. A process for the use of refinery gases made a Kentucky synthetic rubber plant obsolete, but the older process, which used ethyl alcohol, was feasible in India, where ethyl alcohol was inexpensive and where refinery gases were not expected to be readily available for some time. Used machinery can be attractive when ventures entail high risk owing to uncertainty of consumer demand, political instability, fear of labor strife, the possibility of shortages of inputs, or any number of other factors. Reducing sunk costs, where feasible, is common sense under these conditions. The fear of changes in the rules of the game by either the United States or Mexico that allows the maquiladora ("twin") plants to operate probably has a good deal to do with the high incidence of used machinery cited earlier (Pronk 1981, p. 46).

Lead time in acquiring equipment (when excessive for new equipment) will render used machinery appealing. Michael Ecksel, publisher of the Industrial Machinery News, reports that in mid-1981 a wait of 24 to 36 months for new metalworking equipment is common, thus explaining a used-to-new machine sales ratio of around seven to one for this type of equipment. A two to one ratio is more compatible with long-run trends.[3] Occasionally the greater divisibility of earlier vintage equipment can reduce the gestation costs of a plant by permitting partial production while the plant construction is incomplete. In the same vein, new equipment often requires expensive atmospheric controls at the plant site, although these costs must be balanced against a smaller plant per unit of output, which is more likely associated with more recent technologies.

Entrepreneurial ingenuity in the form of applying innovations to used machinery is potentially rewarding. Retrofitting old equipment in some cases is economically feasible. For an assortment of innovations on a secondhand paper plant operating in Chihuahua, see James's case study of Papelera de Chihuahua (1973). The plant, consisting of secondhand mechanical-type equipment from Mexico, Germany, and the United States, was automated by Gastone Guglielmina, an engineer by training. Innovations were introduced for safety devices, wells for checking the oil levels, and recycling steam. The most spectacular change, however, that of switching from horizontal to vertical blades for processing the pulp, along with a consequently needed expansion of the rolling cylinders by one-eighth of an inch, resulted in a 60 percent increase in capacity from 25 tons to 40 tons per day! Some entrepreneurs, then, may have a comparative advantage in using secondhand equipment, which is more flexible and amenable to improvement than newer equipment with the latest technologies already built in along with the higher price tag in capital costs.

These special situations do not exhaust the possibilities, but the point is sufficiently clear. A number of elements enter into the calculus of comparing new with used machines that go beyond the orthodox differences in factor costs. Usually one or more of these situations will be a motivating factor when used machinery is employed. The implication is that no inclusive rule of thumb is possible for assessing used machinery as a viable technology; a case-by-case evaluation by potential users of secondhand machinery appears to be the most prudent approach. Judging from market behavior, in the vast majority of cases, new equipment is the superior choice; but in a significant number of situations firms opt for used machinery. Undoubtedly, this is the way it should be, yet the ensuing sections of this chapter will argue that used machinery is suboptimally employed.

## FURTHERANCE OF DEVELOPMENT GOALS

Managers deciding on an appropriate mode of production and focusing on profit maximization as the primary motivation will not usually take aggregate socioeconomic objectives of development into consideration. However, to the extent that careful selection leads to the importation and operation of used machinery, some important goals of economic development can be advanced. Three such possibilities will be discussed: increased efficiency and competition, the beneficial effect on employment, and conservation of foreign exchange.[4]

Prudent producers in developing countries, within reasonable bounds imposed by limited time and availability of information, scan the "technological shelf" and choose the best equipment for the task, given the economic environment in which the choice is made. As was made clear in the preceding discussion, after taking into account factor costs, market size, expected useful life of the equipment, maintenance costs, spare parts availability, and other conditions, firms in developing countries often opt for a method of production that includes used machinery. Unless there are strong suspicions that market shortcomings systematically skew the choice of machinery in favor of used machinery,[5] there is every reason to presume that production is more efficient when used machinery is included as part of the technological shelf than if it were somehow excluded.

The implicit assumption here, however, is that firms are using nominal costs of inputs as a basis for choosing industrial equipment. Often, producing with secondhand machinery will be even more economically efficient than a private calculation will discern. Conventional wisdom of development economics holds that, owing to resource market imperfections not fully understood, some key factor prices in developing countries do not accurately reflect the real opportunity cost of employing them. Specifically, wages for lower skilled labor are higher than would be the case in a Pareto optimal world, while the nominal costs of capital and foreign exchange usually exceed their opportunity costs in developing countries. Note that to the extent that used machinery has higher labor costs and lower capital costs per unit of output (when compared with new equipment), all three of these resource market defects will bias the choice of production technique in favor of newer capital goods. By the same token, once used machinery is selected as the most efficient method based on nominal market costs of factors, social efficiency will probably be even more pronounced.

As an illustration, referring to the cost matrix developed for 1960-vintage cotton textile used machinery, when a 100 percent rise in variable maintenance costs is assumed and capital costs are reduced 75 percent when secondhand equipment is used, cost per meter of cloth was calculated at 18.614 cents. If these hypothetical used machinery costs are "shadow priced" by assuming nominal wages of unskilled and semiskilled workers are actually 50 percent and 75 percent of their social opportunity cost, respectively, and raising the interest rate by two percentage points (from 12 percent to 14 percent) to reflect a higher social opportunity cost, unit costs of production fell to 17.843 cents per meter of cloth.

Two further short comments are in order with respect to efficiency and competition, although brevity belies their importance. The entrepreneurial corps is broadened in developing countries when used machinery is a viable option because of lower capital costs (that is, some marginal entrepreneurs can "sneak under the wire" of a budget restraint). In addition the number of competitors is likely to be expanded because of the smaller feasible scale of earlier vintage technology.

Turning to productively absorbing employment, using secondhand equipment can contribute for three reasons. First, more machines can be purchased with a given amount of investment funds. Second, because used machinery has embodied earlier technology, it is likely to be associated with a higher labor-capital ratio. Third, while not always true, used machinery will typically induce a greater substitution of labor-intensive maintenance activities.

The net impact of used machinery on developing-country foreign exchange supplies should be benign. Starting with the most obvious point, increasing productive efficiency will enhance the competitiveness of export goods and import-competing goods. Also, relative to new machinery, with rare exceptions, annual capital costs per unit of output must be lower in order for used machinery to be a viable option. Reduced capital costs per unit of output can arise from a lower initial purchase price, a better fit to market size, a reduction in the need for special atmospheric control equipment, more flexibility in use (for example, multiple-purpose equipment), and other factors. Thus, the foreign exchange expenditure on capital equipment associated with a given amount of output will be lower for used machinery.

The net balance-of-payments effect for maintenance and spare parts for used machinery is more ambiguous, but it would appear that a net saving would occur. On the one hand, used machinery will necessitate more spare parts. On the positive side, however, the less complex nature of the parts and the higher volume of demand for them mean that more domestic fabrication and rebuilding of spare parts take place.[6] A reduction of the need to import expensive and highly skilled technicians to service late vintage equipment represents a large saving.

In conclusion, the careful selection of used machinery by individual firms in developing countries is compatible with the overall objective of economic development. Unfortunately, as hinted above, there is every reason to believe that the employment of used machinery is dampened by a number of conditions existing in developing countries themselves or inherent in the organization of the used machinery.

## MARKET IMPERFECTIONS

The typical divergence in developing countries between nominal costs of labor, capital, and foreign exchange and their social opportunity costs was mentioned above. Sometimes deliberate policies, intended to accomplish other goals, contribute to the spread between private and social costs (for example, supporting an overvalued exchange rate in order to encourage imports of capital, maintaining subsidized rates of interest in order to encourage domestic investment, or legislating minimum wages and fringe benefits for workers for social and humanitarian purposes). Whatever the causes, the imperfections of market prices for inputs should not be taken lightly. Another hypothetical example should prove instructive.

Once again, we can turn to the Economic Commission for Latin America's study of cotton textile equipment. When the 1960-vintage technology was adjusted by raising variable maintenance costs by 50 percent, 100 percent, 150 percent, and 200 percent and capital costs were reduced by 25 percent and 50 percent in order to get a hypothetical matrix of unit costs of production for used machinery, only five of eight combinations of 1960 used machinery were more efficient than the new 1965 equipment. This same matrix of costs was recalculated after "shadow pricing" wages of unskilled workers and semiskilled workers at 50 percent and 75 percent of their nominal wages, respectively, and raising the rate of interest from 12 percent in the original study to a range of 13 percent, 14 percent, and 15 percent. The new 1965 equipment was similarly shadow priced. Under these conditions almost all of the resulting permutations for used 1960 equipment (21 of 24 possibilities) were more efficient than the new 1965 machinery.[7] Naturally, no precise measurement of factor price sensitivity for used machinery can be drawn from this exercise, yet the general connotation implies that any constellation of policies resulting in resource costs more nearly reflecting their true economic cost will expand the feasible choices of secondhand capital goods. Conversely, existing conditions tend to dampen the volume of workable used machinery options.

Attitudinal biases intervene as well. Technical experts are inclined to apply what they know best; thus, engineers are prone to use engineering criteria (such as productivity per man hour, production per unit of time, production per unit of energy consumed), which will not necessarily result in lower unit costs of production. The engineering bias is further encouraged by the relative ease of obtaining information on new equipment.[8] Sometimes used machinery is shunned owing to infavoidance, a term used here to mean the desire to avoid the appearance of inferiority. At the microlevel of the individual firm, these attitudinal biases probably do not constitute a long-term threat to the employment of used machinery. Individual entrepreneurs and engineers, those most closely associated with the selection of equipment, catch on relatively quickly as to whether secondhand equipment will work or not, either through personal experience or observation of competitors who use it. Far more damage is done when the engineering bias, often through foreign advisors or experts or infavoidance, emerges at the macropolitical level, resulting in discriminatory measures or even prohibitions against using some types of secondhand machinery.

In the typical developing country much of the fixed investment in productive industrial capacity is financed directly or indirectly by domestic government agencies, international lending agencies, or a combination of both. These organizations have tended to shy away from including used machinery in projects they fund, partly because any enterprises involving used machinery are especially vulnerable to political attack. Blame can be placed on the equipment rather than failure to hedge inventories during falling prices, excessive nepotism, poor supervision, the evils of absentee ownership, a sudden shift in demand, or similar conditions that can disadvantage or doom a fledgling project. The wise course of action from the political standpoint is to eschew used machinery even though such risk aversion may not lead to the most efficient use of funds.

Trade restrictions frequently are biased against used machinery. Both the United Nations (1966) and Economic Intelligence Unit (1962) studies found cases of outright embargoes of some types of used machinery in Argentina, India, Iran, Iraq, Somalia, and Turkey. During telephone conversations with machine dealers during the summer of 1981, many dealers specializing in used machinery sales to Mexico complained of the increasing difficulty of getting past Mexican customs during the previous six months. Furthermore, it is common to apply the same specific tariff on both new and used pieces of equipment even though the latter is almost certain to be considerably less valuable. Even in the case of ad valorem tariffs, some customs officials substitute a valuation close to the price of new equipment rather than accepting the actual purchase price.

One condition in developing countries does bias the selection of equipment in the direction of used machinery. When a budget restraint is present, often the case with small-sized or medium-sized firms, secondhand equipment may be selected because new cannot be afforded. This favorable bias is somewhat offset, perhaps, since it is far more difficult to obtain credit on acceptable terms when used machinery is put up as collateral.

Thus far, attention has concentrated on conditions prevalent in developing countries. The used machinery market itself also harbors imperfections. Traditionally the sale of equipment by the original using firm has been treated cavalierly. Getting rid of "the stuff" in the most convenient way, without much effort to maximize the sale price, has been the usual approach. This seems to be altering. Chemical Week (1978) reported that progressive chemical process companies are developing special managers who work to squeeze out the maximum price for serviceable used capital goods. Now known as "investment recovery" specialists, through the use of computerized information on the specifications of the equipment and the timing of its availability for sale, they circulate information within their own company and advertise to other companies and equipment dealers. An Investment Recovery Conference is held by such experts twice a year. These sales can be enough to interest even a large firm. In 1977 Union Carbide's Investment Recovery Department did $26 million of business and those for DuPont and Dow Chemical $10 million and $4 million, respectively.

Another problem characteristic of the used machinery market may be in the process of partial correction. When the author conducted research in the early 1970s, not a single used machinery dealer was found who did over 20 percent of the total business with developing countries. Although the research was not exhaustive, it was extensive enough for a convincing case that there were very few, if any, firms that specialized in dealing with developing countries. This condition was conducive to treating orders from developing countries as marginal and windfall. No incentives existed for investing the time and effort necessary to establish long-term working relationships, mutual trust, and bases for clear communications. Nor were motivations strong for becoming specialists in dealing in foreign languages, preparing and crating for tropical conditions, and mastering the intricacies of customs regulations and variations in the intensity with which they are enforced. This was one of the reasons why, even with perfectly upstanding dealers, purchasers in developing countries have occasionally received the wrong or useless equipment, the right equipment with the wrong spare parts, equipment that rusted while sitting on an open dock during the two weeks it took to get through customs, or all of the equipment except one essential component which was not packed. Such horror stories abound.

Apparently this situation is changing for the better. Telephone interviews with used machinery dealers during the summer of 1981 turned up many firms who conduct 20 percent or more of their business with Latin American customers. It is not unusual for these dealers to advertise in a Spanish language trade journal, Noticias de maquinariá industrial, which is produced and distributed throughout Latin America by the Industrial Machinery News Corporation of Southfield, Michigan. The author was unable to ascertain whether similar specializations are developing among Western European used machinery dealers and purchasers in Africa and Asia.

Unfortunately, but predictably, unscrupulous practices constitute another problem with the used machinery market. As one director of a used machinery dealers association wrote in 1981, "Regretably there are always 'cowboys' in the machinery and other entrepreneurial trades." He went on to point out that the seller does not have a monopoly on shady behavior. "In fairness, we have found that there is just as much deviousness on the buyer's side in many developing countries." All the more reason to applaud the emergence of firms with special expertise in handling business in developing countries.

## SUMMARY, CONCLUSIONS, AND SUGGESTIONS FOR FURTHER RESEARCH

There are sound theoretical reasons why used machinery should be seriously considered by Third World firms, based on factor price differences between developing countries and the mature industrial countries. Yet, judging from cost comparisons, resource cost differentials alone are incapable of explaining the volume of secondhand equipment being transferred to developing nations, which, as an educated guess, ranges upward of $100 million and comprises at least 10 percent of the value of industrial machinery formation in those countries. An assortment of special situations, by complementing the factor price situation, provides much stronger incentives for employing used machinery. No one of these unusual situations is of overwhelming importance, but collectively they heavily influence the developing-country demand for used machinery. Furthermore, to the extent that managers in developing countries correctly scan the "technological shelf" and choose to employ some used capital goods, this is compatible with the overall economic objectives of economic growth and development. Lamentably, a medley of frictions, either within developing-country economies or embodied in the used machinery market organization, are almost certainly leading to a suboptimal demand for used machinery on the part of the Third World.

This is where the analysis stands, but before going further it behooves us to mention briefly three items that have been purposely neglected up to this point. First, although much has been said about the developing-country demand for used machinery, little has been said about the supply side. Shortages occur intermittently in individual, specialized pieces of equipment, but since developing-country purchases account for a tiny fraction of total market sales of used machinery, in the aggregate at least, supplies of used machinery do not constitute a problem. Developing countries could double their relative shares of the global secondhand purchases and have only a marginal impact on the used machinery market.

Second, no mention has been made of the "plot syndrome," the ultimate extreme of the political bias against used machinery, which views the sales of used machinery to developing countries as part of the plan to keep the Third World in a perpetual state of technological subservience. There are a variety of rejoinders to this proposition, but we can confine ourselves to the observation that if a plot is indeed afoot, developed countries are doing an incredibly inept job of implementing it. Developed countries sell over 95 percent of used machinery to themselves!

Third, some assessments of used machinery have focused on whether businesspersons in developing countries should purchase used machinery at all, pointing out problems with spare parts, maintenance, shorter expected usefulness of the equipment, difficulty in obtaining guarantees or service contracts, and so on.[9] All of these difficulties do indeed often exist. But to overemphasize their impact on the ultimate profitability and efficiency of producing with used machinery seems unwarranted. Do firms in developing countries purchase tens of millions of dollars of used machinery, and continue to do so year after year, because they are stupid or misled? Could secondhand equipment users continue to compete with a handicap of serious inefficiencies? Or do such firms continually go broke only to be replaced by entrepreneurs eager to do the same? None of this makes sense. Entrepreneurs in developing countries are much smarter than this. The fact is that many individuals carefully appraising their options and choosing equipment accordingly constitute the ultimate assessment of used machinery as a viable technology for developing countries.

What specifically can improve matters? First a number of actions centering on improved information would help. It is an expensive matter to inspect and appraise major pieces of used equipment. A representative fee for equipment appraisers in 1981 is $500 per day plus expenses. Yet, in most cases inspection and appraisal are highly advisable. However, for some cheaper secondhand machinery that is fairly standard with a high volume of sales, a system of grading could be instigated by professional associations. Grading, analogous to the system of rating coins in numismatics, is not a costless activity, but it would be less expensive than a full-fledged inspection while giving sufficient information to a prospective buyer.

An international system of reporting, in advance when possible, the availability of an entire plant for sale would appear advisable. Advance notice would increase the possibility of observing the plant while under power and the pretraining of key personnel. The roster of available plants could be maintained by the United Nations Industrial Development Organization or perhaps by a private firm operating a full range of brokering services.

Actually a tremendous amount of information on the availability of used machinery already does exist in a cheap, reliable form. The Used Equipment Directory, Locator of Used Machinery and Equipment, and Industrial Machinery News are three examples of trade publications produced in the United States that contain brief descriptions of countless machine tools and other types of used machinery available for sale. There is no problem of having to generate information "from scratch," but firms in developing countries need to be aware that this information exists. Subscriptions by major industrial parks and branches of industrial development agencies in developing countries could increase the exposure.

As for governmental policies within developing countries, trade policies should be adjusted so there is no discrimination against used machinery vis-à-vis new machinery. If there is fear that used machinery is being overpriced in order to circumvent foreign exchange restrictions, a review board can be established to determine a fair market value for the equipment. Also, domestic agencies as well as international lending entities should at least consider objectively whether used machinery should be included in developing projects. Above all, policy makers and technical decision makers at the national level must be exposed to convincing, concrete evidence of the potential contribution of used machinery to industrial development goals. Since much remains to be done in developing the full range of evidence, we are led naturally to the matter of further research.

Having been forced to rely on hypothetical exercises, I am acutely aware of the paucity of hard data. If the reader became weary of digesting hypothetical figures, be assured that I was tired of using them. Two efforts to improve the situation are the following: first, a study of the aggregate international trade in used machinery using such means as surveys of equipment dealers, multinational corporations, and equipment importers and sample of import records; second, a series of case studies comparing firms that routinely employ used machinery with those that do not. It is also advisable to investigate in a detailed manner firms or projects that have failed when operating with secondhand capital equipment. It is very likely, especially for a global survey, that the prestige and backing of an international organization will be necessary in order to obtain the requisite cooperation from individual firms and government agencies.

Finally, if we know too little about the trade in used industrial machinery, we know even less about the transfer of secondhand agricultural and construction equipment to developing countries. Some investigations here should prove fruitful.

Once a friend who works with an international organization remarked, "But Dilmus, used machinery is not a path to development." He was absolutely correct, of course. Yet, doing a lot of little things well can be a path to development, and the optimal employment of used machinery is one of them. If our estimates are close, used machinery is marginal to the industrial capital formation of developing countries, but the margin is not insignificant. It is marginal in the sense that an arm is marginal to the body or the automobile industry is marginal to the U.S. economy. Given its relative importance, the author is mystified as to why used machinery has remained in a state of relative neglect. Compare the paucity of literature reviewed earlier in this essay with the uncanny explosion of books, documents, pamphlets, journal articles, and newsletters dealing with appropriate technology. Indeed, there are now many directories of organizations dealing with appropriate technology. Surely the investigation of used machinery is a legitimate subhead of appropriate technology, but apparently it has too much of a "blue-collar" ring for the dilettante and lacks a sufficient aura of romance for the fervent true believer. Heroic though such an aspiration may be, the intent of this chapter is to heighten the interest of both orthodox investigators and followers of the appropriate technology movement in scrutinizing the current and potential contribution of used machinery to Third World industrial development.

## NOTES

1. A New England used machinery dealer told the author that in 1980 his firm arranged the sale of a $35 million secondhand plant to Mexico.

2. For the complete matrix of adjusted unit costs, see James (1974: 186-87).

3. Note here that numbers of pieces of equipment are being considered, not dollar volume.

4. These are the most important aggregate benefits. For the possible effect of employing used machinery on saving capital used in anticipating market growth and hastening worker learning, as well as its use for training, rural development programs, and as prototypes for new technologies, see James (1974).

5. Precisely the opposite is actually the case, as the following section explains.

6. Virtually every plant employing used machinery in developing countries that the author has studied fabricated and repaired some spare parts at the plant site, had the work done locally, or some combination of both.

7. The complete data is found in James (1974).

8. The search for information and the necessary learning and problem-solving in locating, inspecting, packing, transporting, installing, and adjusting used machinery can be time-consuming and expensive. Presumably, prospective purchasers of used machinery attempt to take these possibilities into account when picking a production technique. Thus, this does not constitute a "market bias" in the true sense of the word. If the availability or awareness of the availability of information is suboptimal, however, the results are the same.

9. The author is reminded of an experience in Chihuahua City, Mexico, when making a presentation concerning the economic feasibility of employing used machinery at the request of the government of the State of Chihuahua. During the subsequent discussion period, one local businessman brought up the issues of maintenance, spare parts availability, technological obsolescence, and a raft of other real and supposed problem spots for those producing with used machinery. While conversing later at a social gathering, the author was surprised to find that the vociferous gentleman consistently employed a large proportion of used machinery in his forestry operations located in another part of Chihuahua. When I asked him why, his terse reply was, "I can make more money that way."

## REFERENCES

Bhalla, Ajit S. 1975. Technology and Employment in Industry. Geneva: International Labour Office.

Cooper, Charles, Raphael Kapinsky, and R. Turner. 1975. Second-hand Equipment in a Developing Country: A Study of Jute Processing in Kenya. Geneva: International Labour Office.

Economic Commission for Latin America. 1966. "Choice of Technologies in the Latin American Textile Industry." Mimeographed. Santiago, Chile: ECLA.

Economic Intelligence Unit. 1962. American Industry's Potential for Providing Used Machinery and Technical Assistance for Developing Countries. New York: EIU.

International Labour Office. 1968. "The Use of Second-Hand Equipment for Vocational Training Projects." Mimeographed. Geneva: ILO.

James, Dilmus D. 1975. "Second-hand Machinery in Development: A Comment." Journal of Development Studies 11: 230-33.

_____. 1974. Used Machinery and Economic Development. MSU International Business and Economic Studies. East Lansing: Graduate School of Business Administration, Michigan State University.

_____. 1973. "Used Automated Plants in Less Developed Countries: A Case Study of a Mexican Firm." Inter-American Economic Affairs 27: 31-46.

Moore, Frederick T. 1961. "Economic Growth and Foreign Aid: A Proposal Concerning the Export of Industrial Plant." Mimeographed. Santa Monica, Calif.: Rand Corporation.

Netherlands Economic Institute. 1958. Second-hand Machines and Economic Development. Rotterdam: NEI.

Pack, Howard. 1978. "The Optimality of Used Equipment: Calculations for the Cotton Textile Industry." Economic Development and Cultural Change 26: 307-25.

Pronk, Maarten L. 1981. "Offshore Industries and Economic Development: The Mexican Experience." Vakgruep Sociale Geografie van Ontwikkclingslanden. Discussion Paper no. 10. Utrecht, Netherlands: University of Utrecht.

Ralph M. Parsons Company. 1965. Used Equipment Study. Final Report for Agency for International Development. Los Angeles and New York: Parsons.

Schonfield, Andrew. 1962. The Attack on World Poverty. New York: Random House.

Schwartz, Sandra L. 1973. "Second-hand Machinery in Development or How to Recognize a Bargain." Journal of Development Studies 9: 544-55.

Sen, Amartya Kumar. 1962. "On the Usefulness of Used Machines." Review of Economics and Statistics 44: 346-48.

Smith, M. A. M. 1975. "International Trade in Second-hand Machines." Journal of Development Economics 1: 261-78.

Strassmann, W. Paul. 1968. Technological Change and Economic Development: The Manufacturing Experience of Mexico and Puerto Rico. New York: Cornell University Press.

"They Squeeze Dollars Out of Used Equipment." 1978. Chemical Week 123 (October): 31.

Waterson, Albert. 1964. "Good Enough for Developing Countries?" Finance and Development 1: 89-96.

United Nations. 1966. Report of Expert Group on Second-hand Equipment for Developing Countries, 7-22 December 1965. New York: UN.

Vaitsos, Constantine V. 1970. "Transfer of Monopoly Rents." Economic Development Report no. 168. Mimeographed. Cambridge, Mass.: Center for International Affairs, Harvard University.

"Why Used Equipment Is Selling Again." 1977. Modern Plastics 54 (March): 58-60.

# 13

# MANAGING TECHNICAL COOPERATION: THE ROLE OF THE UNIVERSITY

MAURICE I. MIDDLEBERG
FRANCIS J. BARROS

## DEPENDENCY AND TECHNICAL COOPERATION

If there is a prevailing theme in the chapters of this text, it is the ubiquity of dependence. This condition of dependence is manifested in an endless variety of forms: the need to import equipment and expertise, the paucity of local educational institutions, the difficulty Third World scientists have in achieving recognition in the international scientific community, the continuing brain drain, the orientation of Third World scientists to the Western rather than local scientific agenda, ad infinitum. To take but a few concrete illustrations from the current readings, the increased agricultural productivity reported by Atwal was based upon seed varieties developed by Western scientists (see Chapter 9); the used machinery studied by James is produced almost exclusively in developed nations (see Chapter 12); and Lamberton reports that Western sources dominate the transmission of even those telecommunications that are intended to serve a Third World region (see Chapter 11). Moravscik sums up this condition in Chapter 1 when he writes, "Intentionally or unintentionally, the countries with the know-how and knowledge will dictate the pace of the world and the other countries will have little choice but to participate in this or isolate themselves into stagnation."

The contrary condition, independence, is, of course, a fiction in the scientific enterprise. Practitioners of science and technology are inherently dependent upon the work of their predecessors and peers. What is really at issue is the continuing imbalance in the flow of knowledge. Scientific and technological production is the almost exclusive province of developed nations, with developing nations relegated to the role of consumers. In the long run, the goal of developing nations must be to achieve a competitive position, to serve equally as producers and consumers. Moreover, the building of local science and technology capacity is essential to the production of technologies appropriate to local needs and constraints.

The inherent contradictions and difficulties in building local capacity even in a relatively well-endowed country, such as Mexico, are described by Marqez in Chapter 5. As she points out, policies that encourage essential foreign investment simultaneously crowd out efforts to develop local science and technology capacity. Similarly, efforts to integrate extant R&D capacity with the local productive system have proved markedly unsuccessful. These conditions prevail in the Third World and are exacerbated in low-income countries. Hetman aptly summarizes the short-range to midrange future in Chapter 2, "Since technological innovation is concentrated in industrialized countries, technology policy in developing countries can bear mainly on the choice of imported technologies and their optimal utilization."

If, as we believe, Hetman's assessment is correct, then the question is one of the most effective technology transfer strategy. The dilemma is that the transfer strategy must concurrently serve three masters. First, it must be responsive to the policy framework. Weaver and Jameson have elucidated the alternative paradigms—capitalist, socialist, and growth with equity—that underlie development strategies.[1] Each paradigm has very different implications for the type of technology deemed most desirable. The capitalist model places almost exclusive emphasis on fostering economic growth, with relatively little attention given to equity or dependency issues. Increasing productivity is, therefore, of primary importance. In practice, this has meant that relatively sophisticated, capital-intensive technologies are the most highly valued. At the other extreme are the growth with equity exponents who argue that increased productivity can and must be associated with a concern for meeting the basic needs of the poor. This means that labor intensive or other appropriate technologies are intrinsically preferable. These basic models, in turn, guide policy formulation in specific sectors (for example, small farmer techniques are deemed desirable [or undesirable] as opposed to those that presume large holdings and mechanization).

On a second level, the strategy must be responsive to the implementing organizations. It is one thing for top policy makers to announce a preference for labor-intensive technologies. It is quite another matter for the institutions that are supposed to serve as intermediaries in the transfer process (local universities, R&D institutes, extension services, and so on) to implement such a program. Different kinds of expert services, equipment, and organizational structures will be required under each paradigm.

Finally, the transfer strategy must be responsive to the end users of technology. This point hardly needs belaboring. There are endless incidents of developing nations importing technologies that are simply useless to the final recipients. Such incidents arise because those responsible for technology transfer make unwarranted assumptions about the environment in which the technology will be employed.

The pages that follow address a small but critical element of the transfer strategy—technical cooperation. In particular, they shall be concerned with how an external agent (specifically, the U.S. university) can play a constructive role. They shall argue that the core problems in technical cooperation are managerial in nature. Finally, a suggestion is made that a management approach will facilitate responsiveness at all levels while simultaneously serving the critical function of building linkages between levels.

As a final introductory note, the context for the discussion should be set. It will be concerned only with projects or programs that take place under the sponsorship of a Third World government, whether or not the source of funds is an external development assistance agency. Such programs dictate a responsiveness to government concerns not called for in other forms of university-Third World interchange. Collaboration between scholars, the education of students who are not government sponsored, and other independent forms of scientific exchange are entirely different in nature. The attempt to intervene in such independent exchanges treads upon properly sacrosanct academic and educational freedoms and raises questions beyond the scope of this chapter. It is concerned only with a restricted class of activities—Third World government-sponsored technical cooperation projects.

## THE ROLE OF THE UNIVERSITY

The preceding discussion has outlined the multiple challenges facing external agents attempting to play a useful role in development. The focus of this discussion is on the responsibilities of the Western, and more particularly the U.S., university in fostering development, in part because it is the agency with which we are most intimately familiar. More important, the Western university has come to play a critical role in developing and applying the human resources essential to the building of science and technology capability. The problems bedeviling the university are illustrative of the tensions in even well-intentioned efforts at North-South technical cooperation.

From the perspective of a Third World nation laboring under a scarcity of skilled personnel, the rationale for university involvement is straightforward. The Western university is a "knowledge factory"; it offers an immense reservoir of expertise that can be ostensibly applied to a wide variety of needs. For the U.S. university the motives for involvement range from the noble to the prosaic, representing an admixture of genuine commitment to Third World concerns; the intellectual predilections of faculty and staff; the desire to create or modify promising academic programs; the presumed prestige of international involvement; and, not least, the hope of profiting financially by attracting foreign students or through contract work.

The absolute magnitude of cooperation between the U.S. university and the Third World is difficult to judge, since the forms of interchange are so diversified. Nevertheless, some figures are demonstrative of the dimensions of the issue. For the academic year 1979/80, approximately 227,000 Third World-origin students were enrolled in U.S. universities.[2] Based on 1976 data, approximately 22 percent of this total were enrolled in engineering and 12 percent in science programs.[3]

Universities have also served as a major resource for the U.S. Agency for International Development (AID). For the fiscal year 1980, 131 universities held 349 AID contracts and grants valued in excess of $261 million. Of this total, approximately $63 million was for direct technical assistance to developing nations, $10 million for training, and $66 million for research, with the remainder for various technical and support services.[4] In addition to AID, universities have been major recipients of development-oriented contracts and grants from other U.S. government agencies, such as the Department of Agriculture and the National Science Foundation, international organizations, foundations, and Third World governments.[5]

The depth of university involvement is exemplified in the evolution of an elaborate network of professional associations, such as the Association of U.S. University Directors of International Agricultural Programs, and consortia, such as the Midwest Universities Consortium for International Activities, the Consortium for International Development, and the Southeastern Consortium for International Development.

The fact of involvement is not, of course, prima facie evidence for the value of university efforts. Nor, as Morgan reports, is there any empirical evidence that would substantiate either a positive or negative judgment.[6] Universities, however, have a positive role to play. Third World nations are correct in their perception of the Western university as a vast intellectual resource, which, if properly applied, can make an important contribution to development. More to the point, the basic tasks of technical cooperation—training, research, and expert consultation—are synonymous with the primary functions of U.S. universities. This triumvirate has been the dominant ethos of the U.S. university for well over a century. Albeit that their exact contribution is unknown, they are perhaps the one institution whose fundamental purposes are akin to the demands of technical cooperation. Moreover, the public service heritage runs deep in U.S. higher education. Universities have, over the past 30 years, demonstrated a willingness to extend the boundaries of service to other nations. Though not without their points of rigidity, universities have shown a remarkable capacity to adapt to changing social and economic needs. Their almost complete autonomy in matters of curriculum and scholarship has engendered a remarkable flexibility. That flexibility has been and can continue to be harnessed to meet the needs of Third World nations.

From the foregoing it is clear that U.S. universities have assumed an important role in building Third World science and technology capabilities.[7] In making this statement, one hastens to emphasize that universities can make only a partial contribution even when their particular functions are properly executed. Higher education has too often been oversold as a panacea by well-meaning advocates. The other elements for the solution to scientific incapacity—resource transfers, local science and economic policies, and so on—are equally and perhaps more important.

## CRITIQUE OF UNIVERSITY PROGRAMS

Despite their potential, the technical cooperation and human resource development programs of U.S. universities have often been subjected to severe criticism. Given the mixed history of technical cooperation, it would be surprising if university programs were immune from deficiencies. Among the complaints most commonly voiced are the following:

1. Educational programs, whether degree or short-term training programs, are irrelevant to the working conditions and needs of Third World nations.
2. Research and institution building programs import technologies and methods that are inappropriate to the local environment.
3. The relationship between universities and their Third World counterparts (be they other universities, government agencies, or private corporations) are characterized by what Morgan terms a patron-client relationship rather than equality and mutuality.[8]
4. The irrelevance of U.S. educational programs contributes to a twofold brain drain. In part, this is the well-documented exodus of Third World scientists and technicians from their home countries. More subtle is an intellectual orientation toward the concerns of the Western intelligentsia and away from more pressing domestic concerns.

5. Technical cooperation programs are peripheral to the central concerns of the universities and hence fail to attract the requisite resources and support.

6. Faculty sent on technical cooperation missions are poorly prepared for their assignments.

7. The value of university implemented science and technology development projects is indeterminate, as systematic evaluations are almost never carried out.

The nature of the criticisms suggests that the central problems are managerial rather than technical. Seen from this perspective the deficiencies lie in the setting of ambiguous or inappropriate objectives, establishing improper decision-making structures, and failing to create feedback and evaluation mechanisms. Universities are notoriously better at teaching good management than exercising it.

The issue of the irrelevance of educational and training programs demonstrates the centrality of management concerns. The question, of course, is, Irrelevant to what? Irrelevancy can be defined only with respect to a specific development context. As K. N. Rao points out, the science and technology capability of developing nations ranges from those that have a relatively well-developed science infrastructure to those in which the infrastructure is at a rudimentary stage.[9] Much finer distinctions can be made with respect to specific disciplines, problem areas, and manpower needs. In addition, developing-country governments are faced with fundamental policy problems, such as the potential for creating a technocratic elite, the perpetuation of intellectual dependency, and the geographic misallocation of skilled personnel, which must be taken into account in building the science and technology structure. Educational and training programs are perforce irrelevant when they are designed without due consideration of the scientific, economic, and political context. Irrelevancy, then, stems from a managerial lapse (that is, failing to establish clear objectives derived from specific development needs). By extension, the same critique applies to research and institution building projects.

## AN ALTERNATIVE APPROACH: JOINT MANAGEMENT

A general discussion of the management of university science and technology projects is complicated by the many forms technical cooperation can take. Substantively, science and technology components are often critical in fields as diverse as rural, urban, industrial, and educational development, among others. Methodologically, technical cooperation can include degree programs, short-term training, applied and collaborative research, institution building, and expert advisory services. Historically, U.S. universities have contracted for almost every imaginable type of development project, from village-level applications of agricultural technology to interuniversity cooperation to ministerial-level science policy. The dominant practice, however, has been to work with intermediate agencies. That is, they are more likely to help build an extension service or agricultural research center than work directly with small farmers and more likely to help train engineers than design machinery for developing-country industries. Accordingly, the discussion will focus on relations with intermediate institutions.

A basic management strategy is suggested by the very term technical cooperation. Unlike technical assistance, it implies the existence of structures for reciprocity. Technical assistance connotes a one-way flow of information; hence, the likelihood of misunderstanding development needs. We suggest that the principle of joint management must guide the formulation, implementation, and evaluation of technical cooperation projects. Students and practitioners of rural development have recognized the contribution of participatory decision making to project success. We are suggesting a parallel strategy for university-based science and technology projects.

The essence of joint management strategy is shared power in decision making. This approach has two principal virtues. First, it opens up channels of information on needs, intersectoral dependencies, target group characteristics, and project impacts ignored in more closed systems. Second, the process helps build local management capacity by creating administrative structures and providing hands-on experience to a relatively wide group of project participants.

Science and technology projects will involve at least three and possibly four groups—the leadership of the involved developing-country agenc(ies), the individuals or subjects directly affected by the project, the developed-country university, and in some instances, the end-users of the technology. The argument for including participants from different levels is based on several assumptions. First, that perceived needs and constraints are quite different at each level. Second, that linkages between levels are often nonexistent. Third, that participants at any level can undermine or ignore a program that is not responsive to their needs or may adversely affect their self-interest. This latter point is, of course, where modern organizational theory starts. As Etzioni pointed out, obtaining compliance with organizational goals is the fundamental and most difficult task confronting managers.[10] It is precisely the failure to comprehend and respond to the extant pattern of incentives that has undermined so many well-intentioned programs; hence, the urgency of responding to the needs and constraints perceived at each level.

The problem we are addressing is illustrated by Baranson's study of the Brazilian system. "Despite the proliferation of government sponsored R&D facilities in universities, and other research institutes, these facilities rarely have effective linkages with productive activities."[11] Wionczek generalizes the argument when he writes, "Whatever R&D efforts are made in LDC's locally, sometimes at considerable financial and social cost, they are often largely wasted due to the absence of permanent links between [science and technology] activities and local educational and productive systems."[12] The charges of irrelevancy and inappropriateness stem from the atomized condition of the science and technology system. Developed-country universities operate on the presumption that the expressed needs of the intermediate institution are linked to a wider set of demands. In fact, the contrary condition usually prevails.

Our call for a joint management strategy rests upon the need to construct linkages. A prerequisite to successful science and technology projects is developing structures in which the affected parties can share in decision making. This simultaneously permits a more accurate assessment of science and technology needs and builds linkages. A primary task of external agents (such as developed-country universities) is in organizing linking and decision-making structures.

Gostyla and White, in their study of the Guatemalan Institute for Science and Technology (ICTA), provide a concrete example of the approach being suggested.[13] They point out that in the traditional model of an agricultural R&D center, information flows in only one direction. Scientists develop high-yield technologies on an experimental station for transmission to farmers via extension agents. The consistent failure of farmers to adopt the technologies thus developed has been habitually attributed to peasant irrationality or passivity.

The ICTA operates on a contrary set of assumptions (that is, that small farmers are quite rational but that the technologies heretofore developed on tightly controlled experimental stations were simply inappropriate to the real world constraints and incentives facing farmers). The ICTA's success has been attributed to three factors: first, a policy decision that granted ICTA a large measure of autonomy to experiment with alternative R&D models; second, a commitment to use farmers as partners and consultants in the research process (this latter step, when combined with rapid on-site surveys, permitted a much more accurate assessment of the physical and economic limitations facing small farmers); third, a reorganization of the ICTA from monocrop to multidisciplinary research teams. Prior to this, reorganization teams consisted almost solely of experts on a single crop. Farmers, however, grew multiple crops, thereby inhibiting the adoption of technologies that assumed monocropping. The new structure consisted of individuals drawn from various crop specialties plus social scientists, who provided essential socioeconomic data. In short, the ICTA's successes are attributable to organizational adaptations, including a policy decision to decentralize, a restructuring of the R&D center's work processes, and the involvement of the technology users. It is precisely this coordinated involvement of actors at different levels that we have termed joint management.

In practical terms, then, how can developed-country universities go about organizing joint management structures? One must first assume that the general goals and boundaries of the project have been established (for example, building the capacity of an agricultural research center, upgrading an engineering school, developing an industrial extension service). Assuming such a goal has been established, the following are the major elements of the joint management strategy:

1. Identify the affected parties: For which groups or organizations are the goals of the project most salient? This will vary with the nature of the project. In the case of the agricultural research center, it will include the Ministry of Agriculture, the center staff and leadership, and extension agents and farmers. For an engineering school it will mean the responsible unit of the Ministry of Education, the faculty and administration, and the industries that are expected to employ the students and/or services of the school. The more "applied" the intended project outcomes, the more they tend toward the technology end of the science and technology continuum, the greater the need to include technology end users.

2. Build a task-oriented, representative organization involving the affected parties: The question of representation poses some ticklish problems, as the groups most alienated from the target institution are the ones least likely to participate. Moreover, there will always be uncertainty as to whether the designated representative in fact espouses the interests of the particular group in question. There are no easy solutions to the dilemma of representation, but the key is to build on extant organizations. Farmers' groups, industry councils, trade federations, or professional associations usually exist that will represent the user end of the linkage chain. In some cases, special efforts will have to be made to organize user associations and representation. Representation from the target institution and the government will usually, but not always, be easier since the involved parties are more easily identifiable.

Fostering participation requires recognition of two critical factors. First, participation (and organizing for participation) is not a free good. Participation has costs, including staff time, participant time, transportation, organizing meetings, and so on. These costs should be explicitly recognized and provided for in the budget. Second, participation will continue only to the extent that project outcomes remain salient for the participants; this in turn requires the power to direct outcomes.

3. Involve participants in program planning and design: The primary purpose of the joint management strategy is to induce linkages. Accordingly, a necessary step is an assessment of the needs of each party in terms of the overall project goal. This should be followed by the development of objectives that are simultaneously responsive to the needs and constraints of each group. Defining strategies that concurrently meet the interests and limitations of several groups admittedly will often be a difficult task. However, linkages can exist only in the presence of mutual interests. Programs must be designed to strengthen and expand the points of mutuality. The most important management role an external agent can play is to facilitate the discovery of common interests and the formulation of responsive programs. The only alternative is retention of the present atomized system. The greater the extent of participant involvement in program planning and design, the greater the likelihood of identifying common needs and, hence, building linkages.

A related planning issue is organizational adaptation. Newly trained scientists and engineers must have meaningful jobs; researchers must have equipment; extension workers will need transportation; and linkages must be institutionalized. In short, immediate project outputs must exist in a supportive infrastructure. The planning process must also specify what changes in organizational resources and changes are needed to sustain the effects of the project.

4. Involve participants in project implementation: In particular, there should be involvement in project monitoring and evaluation. The presence of a feedback mechanism is critical. When things go wrong (as they undoubtedly will), the organization must be capable of, as Korten puts it, "embracing error."[14] Failures must be seen as opportunities for learning rather than as marks of managerial incompetence. It is absolutely essential to know if linkage-building programs are succeeding. If not, then alternative strategies will have to be devised. Note that a distinction must be made between ex post evaluations and project monitoring. The former is more rigorous methodologically but takes a relatively long time. Implementation monitoring requires rather simple techniques that permit rapid feedback.

## CONCLUSION

Science and technology programs inherently assume a particular development model. External agents, such as Western universities, are often charged with implementing irrelevant or inappropriate programs. In general, such criticisms stem from the failure of the programs to be properly responsive to the local development context. More particularly, science and technology development projects are, at best, responsive only to the needs of the target institution or agency. This narrow focus assumes that the target institution is linked to a wider set of development needs. However, this assumption is usually unwarranted. Accordingly, a joint management strategy is proposed that would simultaneously involve representatives from the policy-making level, the target institution, and the technology end users. This strategy has the dual advantages of increasing responsiveness to local development needs and building linkages.

## NOTES

1. J. Weaver and K. Jameson, Economic Development: Competing Paradigms, Competing Parables (Washington, D.C.: Agency for International Development, 1978).

2. U.S., Bureau of the Census, Statistical Abstract of the United States, 1980 (Washington, D.C., 1980).

3. Ibid.

4. Agency for International Development, "AID Financed University Contracts and Grants" (Active during period October 1, 1979, through September 30, 1980), mimeographed (Washington, D.C.: AID, undated).

5. We leave aside here the use of faculty as individual consultants; our concern is with the institutional role in building science and technology capacity.

6. R. P. Morgan, Science and Technology for Development: The Role of U.S. Universities (New York: Pergamon Press, 1979).

7. We define science somewhat more liberally than does Moravscik, including herein the natural, information, and social sciences.

8. Morgan, Science and Technology.

9. K. N. Rao, "University Based Science and Technology for Development: New Patterns of International Aid," Impact of Science on Society 28 (1978): 117-25.

10. A. Etzioni, A Comparative Analysis of Complex Organizations (New York: Free Press, 1975).

11. J. Baranson, North-South Technology Transfer: Financing and Institution Building (Mt. Airy, Md.: Lomond, 1981), p. 24.

12. M. Wionczek, "Science and Technology Planning," in Integration of Science and Technology with Development/Caribbean and Latin American Problems in the Context of the United Nations Conference on Science and Technology for Development, ed. D. Babatunde and M. Wionczek, pp. 167-68 (New York: Pergamon Press, 1979).

13. L. Gostyla and W. F. Whyte, ICTA in Guatemala: The Evolution of a New Model for Agricultural Research and Development (Ithaca, N.Y.: Center for International Studies, Cornell University, 1980).

14. D. Korten, "Community Organization and Rural Development: A Learning Process Approach," Public Administration Review 40 (September-October 1980): 480-503.

# ABOUT THE EDITOR AND CONTRIBUTORS

MANGALAM SRINIVASAN is currently Director of Technology Policy and Assistance Council at the Center for International Technical Cooperation of the American University in Washington, D. C. Additionally, she is Adjunct Professor of Cybernetics, Science, Technology and Society, and Technology Assessment. She has held the distinguished invitational appointment as a Fellow at the Center for International Affairs at Harvard University, has served as a consultant to the United Nations and its specialized agencies; to the U. S. Department of Energy, the University of California-Berkeley, the U. S. Department of Agriculture, and the Arthur D. Little Company; and has advised science-policy units of many countries, the OECD, the EED, and the Organization of American States at the highest levels. Her numerous publications have addressed the issues of science, technology policy in development, global issues of science, and women in development, among others. She is also the Founder of the International Council for the Responsible Use of Science and Technology.

MICHEL A. AMSALEM is an Assistant Professor at the Graduate School of Business of Columbia University. He previously served as the Economist in charge of Africa at the International Finance Corporation, the organization of the World Bank Group responsible for assistance to the private sector. He is the author of several articles on technology issues in developing countries as well as a book, Technology Choice in Developing Countries (MIT Press, 1982).

AVTAR S. ATWAL is currently Dean of Postgraduate Studies at Punjab Agricultural University in Punjab, India. He is the Founder-President of the Indian Ecological Society, a Fellow of the Entomological Society of India, and a member of numerous international commissions and societies in the areas of pathology, ecology, and entomology, among others. Professor Atwal has authored three books: Applied Animal Ecology (1974), Agricultural Pests of India and South-East Asia (1976), and Agricultural Resources and Production Strategy (1979).

FRANCIS J. BARROS is Vice-Provost for Research and International Development and Director of the Center for International Technical Cooperation at the American University, Washington, D. C. Dr. Barros received his early education in Hong Kong and in Australia and has had extensive experience throughout the world in central project management.

MOMIR CECEZ is a Professor of Economics at Sarajevo University, Yugoslavia, and specializes in economic self-management systems. Professor Cecez worked at the Federal Economic Council, was general manager of a hydropower system, was Secretary for General Economic Affairs in Bosnia and Herzegovina, and was Director of the Institute for Organization, Management and Productivity. He is currently President of the Economic Council of Bosnia and Herzegovina. Professor Cecez worked for the United Nations in Indonesia and Iran from 1968 till 1971. He has written numerous papers and articles on self-management, economic development, and investment, and his last book, Self-Managing Economic System of Yugoslavia, was published in 1980.

FRANÇOIS HETMAN is a Principal Administrator with the OECD Directorate for Science, Technology and Industry. He is author of several books—among which are Europe of Abundance, Secrets of the American Giants, The Language of Forecasting, Mastering the Future, Society and the Assessment of Technology, Science Policy Formats, and Methods of Technological Forecasting—and a number of articles.

DILMUS D. JAMES is a Professor of Economics at the University of Texas, at El Paso. He has written numerous articles about economic, scientific, and technological development of Third World countries, with an emphasis on Latin America and particularly Mexico. His work has been published in the Latin American Research Review, the Journal of Developing Areas, the Western Political Quarterly, and Growth and Change, among others. Dr. James coedited the book Technological Progress in Latin America: Prospects for Overcoming Dependency in 1979.

DONALD M. LAMBERTON has been a Professor of Economics at the University of Queensland, Australia, since 1972. He formerly taught at Case Western Reserve University and Stanford University. In 1978 he was a Visiting Fellow at the Oxford Centre for Management Studies. He authored Share Price Indices in Australia (1958), Science, Technology and the Australian Economy (1970), and The Theory of Profit (1965), which has also been published in Portuguese and Italian. Professor Lamberton is a member of the OECD Group of Experts on Economic Analysis of Information Activities and the Role of Electronics and Telecommunication Technologies and the Australian government's Industrial Property Advisory Committee. Additionally, he is a consultant to the OECD, UNESCO, and the East-West Center Communication Institute Workshop on Economics of Communication.

VIVIANE B. MARQUEZ is currently Associate Professor of Sociology at El Colegio de México and Coordinator of the Program of Science and Technology at that institution. After working for several years on problems of technological dependency and management in the industrial sector of Mexico, she is currently engaged in the field of health, as senior researcher in the Ford Foundation/IDCR project "Bureaucracy and Society: The Case of Public Health in Mexico."

MAURICE I. MIDDLEBERG is Associate Director of the Center for International Technical Cooperation at the American University, having previously held positions at the University of Arizona and Florida State University. He is the author of a number of publications in the field of higher education, administration, and curriculum.

MICHAEL J. MORAVCSIK is a Professor of Physics at the University of Oregon and previously served as Director of the Institute for Theoretical Physics. Educated in Hungary and the United States, Professor Moravcsik also served as Head of the Elementary Particle and Nuclear Theory Group at the

Lawrence Radiation Laboratory, University of California, Livermore. He has been a visiting professor at Purdue University, Scottish Universities Summer School, Harvard University, Oxford University, and Sussex University. He has written over 200 articles dealing with nuclear and elementary particle physics and science policy and organization, particularly in the context of developing countries.

PAOLO SOLERI is Director of the Cosanti Foundation in Scottsdale, Arizona, and is a Distinguished Visiting Lecturer in the College of Architecture, Arizona State University. In 1970 in central Arizona, he began the construction of Arcosanti, a futuristic self-contained urban complex; the social, economic, and ecological implications of its architectural framework are continuously being researched. Major articles on his work have appeared in many publications, including the New York Times, the Washington Post, and Newsweek. He is author of three books: City in the Image of Man, The Sketchbooks of Paolo Soleri, and Matter Becoming Spirit.

VALENTINA NIKOLAYEVA TERESHKOVA is known throughout the world for her space flight as the first woman Pilot-Cosmonaut. Madam Tereshkova made a space flight aboard the Vostok VI and made 48 orbits in 72 hours. Valentina Tereshkova currently serves as President of the Soviet Women's Committee and Vice-President of the Women's International Democratic Federation. She is a Colonel in the Soviet Air Force, a Deputy of the USSR Supreme Soviet, a Member of the Presidium of the Supreme Soviet, and a Member of the Central Committee of the Communist Party of the Soviet Union. She has traveled extensively throughout the world, leading delegations of Soviet women to international congresses, symposia, and seminars, with particular emphasis on children, women's rights, and the problems of peace and security.

CHENG XUEMIN is Chief Engineer of the Bureau of Foreign Affairs, Ministry of Electric Power, in Peking, the People's Republic of China. He is concurrently a member of the Energy Advisory Board of the State Energy Commission in Peking. Xuemin is also the Executive Director of the Chinese Society of Hydroelectric Engineering and has memberships in many scientific and technical bodies. He has had extensive engineering experience both in the United Kingdom and the People's Republic of China and has presented scholarly articles in many international forums, including the coauthorship of "China's Hydro Power Development: Past, Present and Future" for the Water Power 1981 International Conference sponsored by the U.S. Corps of Engineers.